T0179265

SYSTEMS ENGINEERING
Reliability Analysis Using *k*-out-of-*n* Structures

Editors

Mangey Ram

Graphic Era (Deemed to be University), Dehradun, India

Tadashi Dohi

Department of Information Engineering
Hiroshima University, Hiroshima, Japan

CRC Press
Taylor & Francis Group
Boca Raton London New York

CRC Press is an imprint of the
Taylor & Francis Group, an **informa** business
A SCIENCE PUBLISHERS BOOK

CRC Press
Taylor & Francis Group
6000 Broken Sound Parkway NW, Suite 300
Boca Raton, FL 33487-2742

First issued in paperback 2020

ISBN-13: 978-1-138-48292-0 (hbk)
ISBN-13: 978-0-367-78000-5 (pbk)

Library of Congress Cataloging-in-Publication Data

Names: Ram, Mangey, editor. | Dohi, Tadashi, editor.
Title: Systems engineering : reliability analysis using k-out-of-n
  structures / editors, Mangey Ram (Graphic Era (Deemed to be University),
  Dehradun, India), Tadashi Dohi (Department of Information Engineering,
  Hiroshima University, Hiroshima, Japan).
Description: Boca Raton, FL : CRC Press, 2019. | Includes bibliographical
  references and index.
Identifiers: LCCN 2019007547 | ISBN 9781138482920 (hardback)
Subjects: LCSH: Systems engineering. | Failure analysis (Engineering) |
  System failures (Engineering) | Reliability (Engineering)
Classification: LCC TA168 .S888537 2019 | DDC 620.001/171--dc23
LC record available at https://lccn.loc.gov/2019007547

Preface

The main aim of this book is to publish the original research and survey articles that describe the latest research results in k-out-of-n structures, which are useful to improve system dependability. The title "**Systems Engineering: Reliability Analysis Using k-out-of-n Structures**" covers a comprehensive range of k-out-of-n redundancy in various engineering systems. Each chapter was peer-reviewed by at least two referees, where the mathematical analysis, computation techniques/algorithms, tool development, maintenance planning and optimization are considered with industry applications. We believe that this is a unique book which highlights the k-out-of-n structures in reliability modeling, and is beneficial for researchers, graduate students and practitioners to understand the advances in reliability design field.

The following topics are covered in the book:

- Evaluation methods for reliability of consecutive-k systems
- Reliability properties of k-out-of-n:G Systems
- Maintenance policies for k-out-of-n models and their applications to consecutive systems
- Reliability modeling and estimation of (k_1, k_2)-out-of-(n, m) pairs: G balanced systems
- Sequencing optimization for k-out-of-n:G cold-standby systems considering reliability and energy consumption
- Impact of correlated failure on the maintenance of multi-state consecutive 2-out-of-n: failed systems
- Structural and probabilistic examinations of k-out-of-n:G system and their application to an optimal construction of a safety monitoring system

- An analysis of $(k+1)$-out-of-n:F fault tolerant system with fixed warranty period
- Reliability analysis of three-stage weighted 4-out-of-n:F system subject to possibility of degradation after repair with inspection

Mangey Ram
Tadashi Dohi

Contents

List of Contributors

Akshay Kumar, Department of Applied Sciences, Tula's Institute, The Engineering and Management College, Dehradun, Uttarakhand, India, Email: akshaykr1001@gmail.com

Beena Nailwal, Department of Mathematics, Statistics and Computer Science, G. B. Pant University of Agriculture and Technology, Pantnagar, India

Bentolhoda Jafary, University of Massachusetts at Dartmouth, USA; bjafary@umassd.edu

Bhagawati Prasad Joshi, Seemant Institute of Technology, Pithoragarh, India. Email: bpjoshi.13march@gmail.com

E.A. Elsayed, Rutgers University, USA; elsayed@soe.rutgers.edu

Fumio Ohi, Nagoya Institute of Technology, Japan; o-sakae@ob.aitai.ne.jp

Gregory Levitin, The Israel Electric Corporation, Haifa, Israel, gregory.levitin@iec.co.il

Hisashi Yamamoto, Tokyo Metropolitan University, Japan; yamamoto@tmu.ac.jp

Jingbo Guo, Rutgers University, USA

Kodo Ito, Tottori University, Japan; itokodo@sse.tottori-u.ac.jp

Laance Fiondella, University of Massachusetts at Dartmouth, USA; lfiondella@umassd.edu

Lei Zhou, Tokyo Metropolitan University, Japan; 18795900817@163.com

Liudong Xing, Electrical and Computer Engineering Department, University of Massachusetts -Dartmouth, MA liudong.xing@umassd.edu

Mangey Ram, Department of Mathematics, Computer Science & Engineering, Graphic Era (Deemed to be University), Dehradun, Uttarakhand-248002, India

Nupur Goyal, Department of Mathematics, Graphic Era (Deemed to be University), Dehradun, Uttarakhand-248002, India

Prashanthi Boddu, Global Prior Art Inc, Boston, MA, prashanthi85@gmail.com

S.B. Singh, Department of Mathematics, Statistics and Computer Science, G. B. Pant University of Agriculture and Technology, Pantnagar, India

Taishin Naakamura, Tokyo Metropolitan University, Japan; nakamura-taishin@ed.tmu.ac.jp

Tomoaki Akiba, Chiba Institute of Technology, Japan; tomo@akiaki.net

Toshio Nakagawa, Aichi Institute of Technology, Japan; toshio-nakagawa@aitech.ac.jp

Xiao Xiao, Tokyo Metropolitan University, Japan; xiaoxiao@tmu.ac.jp

Evaluation Methods for Reliability of Consecutive-*k* Systems

Tomoaki Akiba[1*], Taishin Nakamura[2], Xiao Xiao[2] and Hisashi Yamamoto[2]

[1] Chiba Institute of Technology, 2-17-1 Tsudanuma, Narashino, Chiba, 275-0016, Japan

[2] Tokyo Metropolitan University, 6-6 Asahigaoka, Hino, Tokyo, 191-0065, Japan

1. Introduction

Consecutive-k system is a general term for the systems, where a concentration of failed components causes the system failure. One of the consecutive-k systems is a consecutive-k-out-of-n system. It has been extensively studied since the early 1980s and was first studied by Kontoleon (1980). The consecutive-k-out-of-n:F (G) systems consist of n components and fail (work) if and only if at least k "consecutive" components fail (work). Based on this definition, we can see that a consecutive-1-out-of-n:F (G) system is equivalent to a series (parallel) system, while a consecutive-n-out-of-n:F (G) system is equivalent to a parallel (series) system.

Consecutive-k-out-of-n systems have also been extended to two-dimensional systems by Salvia and Lasher (1990). Boehme et al. (1992) defined the more general two-dimensional system, which later are called linear or circular connected-$\mathbf{X}$-out-of-(m, n):F lattice systems. The linear connected-$\mathbf{X}$-out-of-(m, n):F lattice system contains the components which are arranged into a rectangular pattern with m rows and n columns. $\mathbf{X}$ denotes the figure consisting of the failed components which leads to the system failure and is called "failure pattern" throughout this chapter. The system fails if and only if all the components which constitute failure pattern $\mathbf{X}$ fail. Salvia and Lasher (1990) also described d-dimensional consecutive-k-out-of-n systems ($d \geq 2$). We call these systems multi-dimensional consecutive-k-out-of-n systems in this chapter.

*Corresponding author: tomo@akiaki.net

One of the most important problems in reliability theory is the reliability evaluation under the assumption that all the component reliabilities are given. In this chapter, we review the evaluation methods for the reliability of consecutive-k systems. In particular, we focus on our previous studies. The reliability of the consecutive-k-out-of-n system can be computed relatively easily via the existing evaluation methods. However, when a multi-dimensional consecutive-k-out-of-n system consists of many components, it takes much computing time to obtain the exact system reliability. To overcome this problem, we also show bounds and approximate values for the reliability of multi-dimensional consecutive-k-out-of-n systems in this chapter. And, we assume the case of independent component failure in the system, through the chapter. Please note that some researchers provided assumption of dependent component failure in the consecutive-k system.

The chapter is organized as follows: First, Section 2 shows the definition of consecutive-k-out-of-n systems and the evaluation methods for the reliability of consecutive-k-out-of-n:F systems. Then, we present the definitions and the evaluation methods for the reliability of connected-X-out-of-(m, n):F lattice systems, and two-dimensional consecutive-k-out-of-n related systems and multi-dimensional consecutive-k-out-of-n systems, in Sections 3 and 4, respectively. Finally, related topics for future research are listed in Section 5.

2. Consecutive-*k*-out-of-*n* systems

In this section, we define consecutive-k-out-of-n systems and show the evaluation methods for the reliability of consecutive-k-out-of-n:F systems. We assume that (1) each component and the system can have only two states, either working or failed, (2) the failures of components are independent, and (3) the component reliabilities are given throughout this chapter.

2.1 Definition of consecutive-*k*-out-of-*n* systems

Consecutive-k-out-of-n:F (G) systems can be classified into linear and circular cases. A linear consecutive-k-out-of-n:F system consists of n linearly arranged components and fails if and only if at least k consecutive components fail (see Fig. 1 (a)). On the other hand, a system in the circular case can be expressed by connecting the components at both ends of the linear consecutive-k-out-of-n: F system (see Fig. 1 (b)).

Let binary variable x_i denote the state of component i ($i = 1, 2, \ldots, n$), and

$$x_i = \begin{cases} 1, & \text{if component } i \text{ fails,} \\ 0, & \text{if component } i \text{ works.} \end{cases} \tag{1}$$

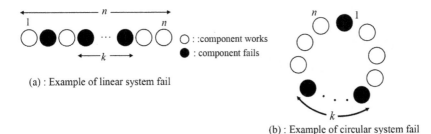

(a) : Example of linear system fail

○ : :component works
● : component fails

(b) : Example of circular system fail

Fig. 1: Consecutive-k-out-of-n:F system

Then, vector **x** represents the states of all components in the system and is known as the component state vector. Let ϕ represent the state of the system, and

$$\phi = \begin{cases} 1, & \text{if the system fails,} \\ 0, & \text{if the system works.} \end{cases} \tag{2}$$

The structure function of a linear consecutive-k-out-of-n:F system is given by

$$\phi^L(\mathbf{x}) = \begin{cases} 1, & \text{if } \sum_{i=1}^{n-k+1} \prod_{j=i}^{i+k-1} x_j \geq 1, \\ 0, & \text{if } \sum_{i=1}^{n-k+1} \prod_{j=i}^{i+k-1} x_j = 0 \end{cases} \tag{3}$$

When we consider the circular cases, $x_i = x_{i-n}$, for all $i = n + 1, n + 2, \ldots, n + k - 1$, for convenience, and the structure function of a circular consecutive-k-out-of-n:F system is given by

$$\phi^C(\mathbf{x}) = \begin{cases} 1, & \text{if } \sum_{i=1}^{n} \prod_{j=i}^{i+k-1} x_j \geq 1, \\ 0, & \text{if } \sum_{i=1}^{n} \prod_{j=i}^{i+k-1} x_j = 0 \end{cases} \tag{4}$$

2.2 Evaluation methods for reliability of consecutive-*k*-out-of-*n* systems

In this subsection, we present the evaluation methods for the reliability of linear and circular consecutive-k-out-of-n:F systems.

For $\phi = L, C$ and $i = 1, 2, \ldots, n$, let $R^\phi(k, i{:}[p_i])$ be the reliability of a consecutive-k-out-of-i:F system with the reliability p_i and failure probability $q_i (= 1 - p_i)$ of component i, where the system is linear (circular) if $\phi = L$ (C), respectively.

Many recursive formulas have been derived in order to evaluate system reliability. Well-known recursive algorithms for evaluating the reliability of consecutive-k-out-of-n:F systems were proposed by Hwang (1982). They are, for $i = 1, 2, \ldots, n$,

$$R^L(k, i: [p_i]) = \begin{cases} 0, & \text{if } i < k, \\ 1 - \prod_{j=1}^{k} q_j, & \text{if } i = k, \\ \sum_{l=n-k+1}^{n} p_l \prod_{j=1}^{n} q_j R^L(k,l:[p_i]), & \text{otherwise,} \end{cases} \tag{5}$$

and,

$$R^C(k, i: [p_i]) = \begin{cases} 0, & \text{if } i < k, \\ 1 - \prod_{j=1}^{k} q_j, & \text{if } i = k, \\ \sum_{l=n}^{n+k-1} p_l \prod_{j=l}^{n+k-1} q_j R^L(k,l:[p_i]), & \text{otherwise,} \end{cases} \tag{6}$$

where $p_i = p_{i-n}$, for all $i = n + 1, n + 2, ..., n + k - 1$ and $\prod_{j=a}^{b} q_j = 1$ for $a > b$. The recursive algorithms, based on the basic idea of Hwang (1982), have been proposed in order to evaluate the reliability of various systems, for example, a linear (circular) r-within-consecutive-k-out-of-n:F system (see Malinowski and Preuss 1995(b); Malinowski and Preuss 1996(b)), similarly, a linear (circular) m-consecutive-k-out-of-n:F system (see Papastavridis 1990; Alevizos et al. 1992), etc.

Fu (1986) proposed the finite Markov chain imbedding approach for evaluating system reliability. Fu and Hu (1987) gave the following $(k - 1) \times (k - 1)$ transition probability matrix, for $i = 1, 2, ..., n$.

$$A(i) = \begin{pmatrix} p_i & q_i & 0 & & 0 \\ p_i & 0 & q_i & \cdots & 0 \\ p_i & 0 & 0 & & 0 \\ \vdots & & & \ddots & \vdots \\ p_i & 0 & 0 & & q_i \\ 0 & 0 & 0 & \cdots & 1 \end{pmatrix}_{(k-1)\times(k-1)}. \tag{7}$$

Then, we can compute the reliability of the linear consecutive-k-out-of-n:F system by using the transition probability matrix $A(i)$ as follows:

$$R^L(k, n: [p_n]) = \pi_0 \prod_{i=1}^{n} A(i) u^T, \tag{8}$$

where $\pi_0 = (1, 0, ..., 0)_{1\times(k-1)}$, u^T is the transpose of the row vector $u = (1, ..., 1, 0)_{1\times(k-1)}$. After that, Fu and Koutras (1994) named this method Finite Markov Chain Imbedding Approach (FMCIA). Zhao and Cui (2009) introduced Accelerated Scan Finite Markov Chain Imbedding (AS-FMCI). AS-FMCI reduces the complexity of computation for system reliability since it decreases the number of the multiplied matrices. Moreover, FMICA has been widely used to evaluate the reliability of various systems, e.g., $< n; f; k >$ systems (see Cui et al. 2006) and consecutive-k_r-out-of-n_r:F linear zigzag structures and circular polygon structures (see Lin et al. 2016). Koutras (1996) and Cui et al. (2010) summarized the results of FMCIA.

For a detailed discussion of consecutive-k-out-of-n systems, the reader is referred to the previous reviews by Chao et al. (1995), Chang et al. (2000), Kuo and Zuo (2002), Eryilmaz (2010), Triantafyllou (2015) and so on. Figure 1 in the Triantafyllou (2015) described an image of the relationship among a consecutive-k-out-of-n and its related systems.

3. Connected-X-out-of-(m, n):F lattice systems

In this section, we define linear or circular connected-X-out-of-(m, n):F lattice systems and show the evaluation methods for the system reliability. "Linear" and "circular" mean the shape of this system. **X** means the figure consisting of the failed components and the system fails if and only if all the components which constitute failure pattern **X** fail. The components in the linear system are arranged into a rectangular pattern with m rows and n columns, as shown in Fig. 2 (a). The components in the circular system are arranged on the intersections of m circles and n rays, as shown in Fig. 2 (b). As an example, when failure pattern **X** represents (r, s), the system (that is, a linear or circular connected-(r, s)-out-of-(m, n):F lattice system) fails whenever there exists a submatrix of r rows and s columns that consists of all failed components. When failure pattern **X** also represents (r_1, s_1)-or-(r_2, s_2), this system is called a linear or circular connected-(r_1, s_1)-or-(r_2, s_2)-out-of-(m, n):F lattice system; it fails whenever there exists a submatrix of "r_1 rows and s_1 columns" or "r_2 rows and s_2 columns" that consists of all failed component sequence.

Salvia and Lasher (1990), Boehme et al. (1992), etc., gave the following examples in order to show where connected-X-out-of-(m, n):F lattice systems may be used:

(1) The presence of a disease is diagnosed by reading an X-ray. Let p be the probability that an individual cell (or other small portions

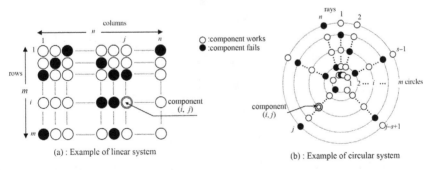

(a) : Example of linear system   (b) : Example of circular system

Fig. 2: Connected-X-out-of-(m, n):F lattice system

of the X-ray) is healthy. Unless diseased cells are aggregated into a sufficiently large pattern (say a $k \times k$ square), the radiologist might not detect their presence. In medical diagnostics, it may be more appropriate to consider a three-dimensional grid in order to calculate the detection probability of patterns in a three-dimensional space (Salvia and Lasher 1990).

(2) A camera surveillance system has 16 cameras arranged into four rows and four columns to monitor a specified area. The distances between two adjacent cameras in the same row or the same column are equal to d. Each camera is able to monitor a disk of radius d. Such a system can be modeled as a linear connected-(2, 1)-or-(1, 2)-out-of-(4, 4):F lattice system (Boehme et al. 1992). The details of this example were considered in Noguchi et al. (1996).

(3) A cylindrical object (e.g., a reactor) is covered by a system of feelers (e.g., sensors) for measuring temperature, as shown in Fig. 3. Each of m circles has n feelers. The system fails whenever a (3, 2)-submatrix of failed components occurs. This system can be modeled as a circular connected-(3, 2)-out-of-(m, n):F lattice system (Boehme et al. 1992).

We define the following notation which is used extensively in this section. Let component (i, j) denote the component located at the i-th row (i-th circle) and the j-th column (j-th ray), with reliability p_{ij} and failure probability $q_{ij} = 1 - p_{ij}$, for $i = 1, 2, \ldots, m$ and $j = 1, 2, \ldots, n$. Letting binary variable x_{ij} denote the state of component (i, j) ($i = 1, 2, \ldots, n$), then

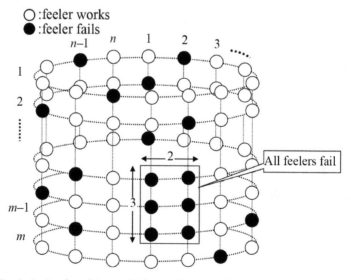

Fig. 3: A circular object, which may be treated as a circular connected-
(3, 2)-out-of-(m, n):F lattice system

$$x_{ij} = \begin{cases} 1, & \text{if component } (i, j) \text{ fails,} \\ 0, & \text{if component } (i, j) \text{ works,} \end{cases} \tag{9}$$

for $i = 1, 2, \ldots, m$ and $j = 1, 2, \ldots, n$.

The evaluation methods for the reliability of multi-dimensional consecutive-k-out-of-n systems are reviewed in Kuo and Zuo (2002). Though they focused on connected-(r, s)-out-of-(m, n):F lattice systems, there are a few descriptions on the other multi-dimensional consecutive-k-out-of-n systems. Accordingly, we also review the evaluation methods for the reliability of the other multi-dimensional consecutive-k-out-of-n systems in Sections 3 and 4.

3.1 Connected-(r, s)-out-of-(m, n):F lattice systems

In this subsection, we consider connected-(r, s)-out-of-(m, n):F lattice systems. The random variable Y_{ij} is defined as

$$Y_{ij} = \prod_{a=i-r+1}^{i} \prod_{b=j-s+1}^{j} x_{ab}, \tag{10}$$

where, in the circular cases, $x_{ij} = x_{i,j-n}$, for all i and $j = n + 1, n + 2, \ldots, n + s - 1$, for convenience. For $\phi = L, C$, let $R^{\phi}(r, s, m, n; [p_{ij}])$ be the reliability of a connected-(r, s)-out-of-(m, n):F lattice system with the reliability p_{ij} of component (i, j) for $i = 1, 2, \ldots, m$ and $j = 1, 2, \ldots, n$, where the system is linear (circular) if $\phi = L(C)$, respectively. Then, the system reliabilities are defined as follows.

$$R^{L}(r, s, m, n; [p_{ij}]) = \Pr\left\{ \bigcap_{i=r}^{m} \bigcap_{j=s}^{n} \{Y_{ij} = 0\} \right\}, \tag{11}$$

and

$$R^{C}(r, s, m, n; [p_{ij}]) = \Pr\left\{ \bigcap_{i=r}^{m} \bigcap_{j=s}^{n+s-1} \{Y_{ij} = 0\} \right\}. \tag{12}$$

When $m = r$ or $n = s$, the connected-(r, s)-out-of-(m, n):F lattice systems can be treated as consecutive-k-out-of-n:F systems (Boehme et al. 1992). For these systems, we can use the evaluation methods for the reliability of the consecutive-k-out-of-n:F systems, e.g. Hwang (1982). For general cases, Zuo (1993) suggested the SDP method. Yamamoto and Miyakawa (1995) developed a recursive algorithm for evaluating the reliability of the linear connected-(r, s)-out-of-(m, n):F lattice system. Since their numerical experiments showed the recursive algorithm to be more efficient in comparison to the SDP method, we present the algorithm of Yamamoto and Miyakawa (1995).

δ_{ij} : index function, which takes 1 when components $(i - r + 1, j)$, $(i - r + 2, j)$, ..., (i, j) all fail, and 0 otherwise.

G : $(m - r + 1)$-dimensional vector, $(g_1, g_2, \ldots, g_{m-r+1})$

B(g_i, i, j) : event that all components fail in the $r \times g_i$ rectangle with component (i, j) as its lower right corner

$R(r, s, m, j; \mathbf{G})$: reliability of a linear connected-(r, s)-out-of-(m, j):F lattice system given that **B**(g_i, i, j) does not occur for each j as shown in Fig. 4.

By using the above notation, we can get

$$R(r, s, m, j; \mathbf{G}) = \begin{cases} \sum_{X_n} R(r, s, m, j; \mathbf{G}') \prod_{i=1}^{m} \left(p_{ij}^{1-x_{ij}} (1-p_{ij})^{x_{ij}} \right) \\ \qquad\qquad \text{if } i \geq \min_{\mathbf{G}} g_i \geq 1, \\ 1, \qquad\quad\;\; \text{if } \min_{\mathbf{G}} g_i \geq i \geq 1, \\ 0, \qquad\quad\;\; \text{if } \min_{\mathbf{G}} g_i = 0, \end{cases} \tag{13}$$

where $\mathbf{G}'$ is a $(m - r + 1)$-dimensional vector $(g'_1, g'_2, ..., g'_{m-r+1})$ and

$$g'_i = \begin{cases} g_i - 1, & \text{if } \delta_{ij} = 1, \\ s, & \text{otherwise.} \end{cases} \tag{14}$$

In addition, $\mathbf{x}_j = (x_{1j}, x_{2j}, ..., x_{mj})$ is the 0-1 binary vector which represents the state of column j and is given by δ_{ij}s. Then, the following equation holds.

$$R^L(r, s, m, n; [p_{ij}]) = R(r, s, m, n; \mathbf{s}) \tag{15}$$

where $\mathbf{s}$ is an $(m - r + 1) \times 1$ vector with all s's.

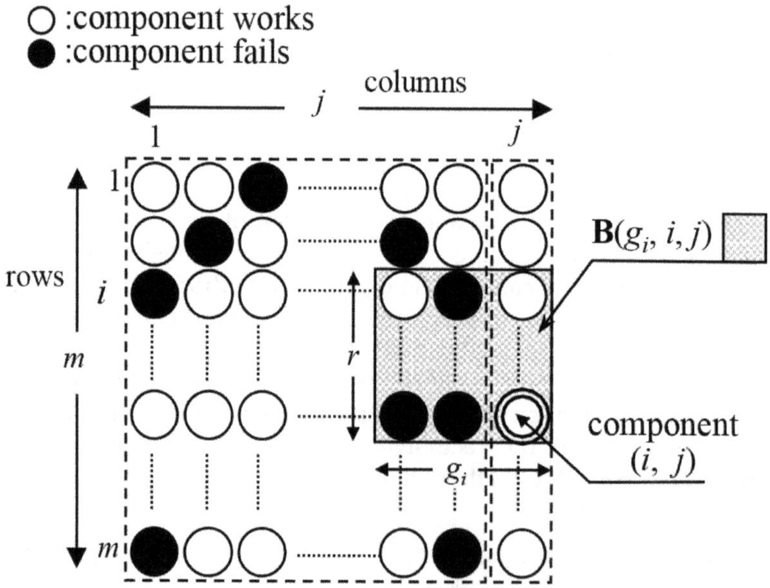

Fig. 4: $R(r, s, m, j; \mathbf{G})$

The total time required for evaluating the system reliability is $O(s^{m-r+1}m^2rn)$ by using the above recursive formula. In other words, its order is exponential for m but linear for r and n.

On the other hand, Zhao et al. (2011) proposed an evaluation method for the reliability of a linear connected-(r, s)-out-of-(m, n):F lattice system by FMCIA. Furthermore, Nakamura et al. (2018) proposed a matrix-based evaluation method for the reliability of linear connected-(r, s)-out-of-(m, n):F lattice systems in the case of $r = m - 1$ and $r = m - 2$.

Yamamoto and Miyakawa (1996(b)) and Yamamoto and Akiba (2005(a)) derived recursive formulas for the reliability of the circular connected-(r, s)-out-of-(m, n):F lattice system. Yamamoto and Akiba (2005(a)) showed that their recursive formula is more effective than that of Yamamoto and Miyakawa (1996(b)) as the number of rays (n) becomes larger.

The upper and lower bounds for the reliability of the linear connected-(r, s)-out-of-(m, n):F lattice systems were reported by many papers (e.g., Yamamoto and Miyakawa 1995; Yamamoto and Miyakawa 1996(b); Malinowski and Preuss 1996(a); Koutras et al. 1997 and Boutsikas and Koutras 2000). Boutsikas and Koutras (2000) indicated that the best available bounds for a linear system were the upper bound U_{FK} of Boutsikas and Koutras (2000) and the lower bound L_{YM} of Yamamoto and Miyakawa (1995) from their numerical experiments. Moreover, Beiu and Daus (2015) reviewed several bounds for the reliability of the linear connected-(r, s)-out-of-(m, n):F lattice systems.

For the circular case, the best upper and lower bounds among the above have not been identified because numerical experiments for all the above bounds have not been executed in the same environment.

We explain a limit theorem in the i.i.d. (independent and identically distributed) case for approximate values of system reliability. For example, we consider the series system that consists of n components with a component reliability $p(0 < p < 1)$. Then, it is obvious that the system reliability is expressed as p^n. If n approaches to infinity, then the system reliability approaches to zero. If p approaches to one, then the system reliability approaches to one. Now, we assume that $p_n = 1 - \lambda/n$, that is, p_n is a function of n, where λ is a positive constant value. When n approaches to infinity (then, p_n approaches to one), the system reliability approaches to $e^{-\lambda}$. Thus, when the system size n becomes larger and component reliability p approaches to one, the system reliability approaches to the value except for one and zero. In this subsection, the theorem like this is called a "limit theorem". This theorem may give us approximate values of system reliability when system size n is large and component reliability p is almost equal to one. Letting $\lambda = n(1 - p)$, the approximate value of the series system reliability is provided by $\exp[-n(1 - p)]$.

The limit theorem for a linear connected-(r, s)-out-of-(m, n):F lattice system was given by Fu and Koutras (1994).

For the mathematical description, we redefine component reliability p_n as a function of n. Letting $f(\cdot)$ be a monotone increasing function, we suppose $m = f(n)$ is the relation among m and n (the notation and assumptions are used throughout this subsection when we consider limit theorems).

Let $R^L(r, s, m, n; p_n)$ be the reliability of a linear connected-(r, s)-out-of-(m, n):F lattice system with component reliability p_n in the i.i.d. case.

Let λ be a positive constant value. If $\lim_{n \to \infty} mn(1 - p_n)^{rs} = \lambda$, then

$$\lim_{n \to \infty} R^L(r, s, m, n; p_n) = \exp[-\lambda]. \tag{16}$$

Limit theorem can be proved easily from Corollary 2.3 in Fu and Koutras (1994).

For a circular system, the similar limit theorem holds as

$$R^L(r, s, m, n; p_n) \le R^C(r, s, m, n; p_n) \le R^L(r, s, m, n + s - 1; p_n). \tag{17}$$

Therefore, from Eqs. (16) and (17), we can get

$$R^L(r, s, m, n; p_n) \approx \exp[-mn(1 - p)^{rs}], \tag{18}$$

$$R^C(r, s, m, n; p_n) \approx \exp[-mn(1 - p)^{rs}]. \tag{19}$$

Fu and Koutras (1994) also proposed the limit theorem in a non i.i.d. (independent but non-identically distributed) case. Koutras and Papastavridis (1993) and Koutras et al. (1993) proposed the limit theorems in another expression. The (properly normalized) system time to failure converges to the Weibull distribution as n approaches to infinity. Let T_n denote the time to failure of a linear or circular connected-(k, k)-out-of-(n, n):F lattice system. Koutras et al. (1993) derived the following limit theorem. Let a be a positive constant value and $q(t) = (\mu t)^a + o(t^a)$. Then,

$$\lim_{n \to \infty} \Pr\left\{ n^{2/(ak^2)} T_n \le t \right\} = 1 - \exp[-(\mu t)^{ak^2}]. \tag{20}$$

Ksir (2000) also proposed a limit theorem similar to Eq. (20).

3.2 Connected-(1, 2)-or-(2, 1)-out-of-(m, n):F lattice systems

In this subsection, we consider connected-$(1, 2)$-or-$(2, 1)$-out-of-(m, n):F lattice systems. This system is a special case of connected-(r_1, s_1)-or-(r_2, s_2)-out-of-(m, n):F lattice systems. For the sake of convenience, in the linear case as shown in Fig. 5, $x_{i0} = 0$ for $i = 1, 2, \ldots, m$ and $x_{0j} = 0$ for $j = 1, 2, \ldots, n$ and in the circular case, $x_{i0} = x_{in}$ for $i = 1, 2, \ldots, m$ and $x_{0j} = 0$ for $j = 1, 2, \ldots,$

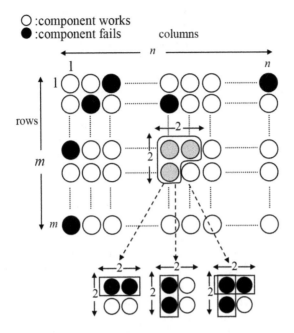

Fig. 5: Example of failure of a linear connected-(1, 2)-or-(2, 1)-out-of-(m, n):F lattice system

n. For $i = 1, 2, \ldots, m$ and $j = 1, 2, \ldots, n$, the random variables $Z_{ij}^{(1)}$ and $Z_{ij}^{(2)}$ are defined as follows:

$$Z_{ij}^{(1)} = x_{i-1,j}\, x_{ij}. \tag{21}$$

$$Z_{ij}^{(2)} = x_{i,j-1}\, x_{ij}. \tag{22}$$

For $\phi = L, C$, let $R^{\phi}(m, n; [p_{ij}])$ be the reliability of connected-(1, 2)-or-(2, 1)-out-of-(m, n):F lattice systems with the reliability p_{ij} of component (i, j) for $i = 1, 2, \ldots, m$ and $j = 1, 2, \ldots, n$, where the system is linear (circular) if $\phi = L(C)$, respectively. Thus, the system reliabilities are defined as

$$R^{L}(m, n; [p_{ij}]) = \Pr\left\{\bigcap_{i=1}^{m}\bigcap_{j=1}^{n}\left\{\left\{Z_{ij}^{(1)} = 0\right\} \cap \left\{Z_{ij}^{(2)} = 0\right\}\right\}\right\}, \tag{23}$$

and

$$R^{C}(m, n; [p_{ij}]) = \Pr\left\{\bigcap_{i=1}^{m}\bigcap_{j=1}^{n}\left\{\left\{Z_{ij}^{(1)} = 0\right\} \cap \left\{Z_{ij}^{(2)} = 0\right\}\right\}\right\}, \tag{24}$$

namely, these systems fail whenever at least two consecutive components fail on a row (circle) or a column (ray).

Many recursive formulas for evaluating the reliability of linear and circular connected-(1, 2)-or-(2, 1)-out-of-(m, n):F lattice systems have been provided in the literature. Yamamoto et al. (2008) proposed the recursive algorithm. It requires $O(n2^{2m})$-time. Furthermore, Nakamura et al. (2017(b)) employed FMCIA in order to evaluate the reliability of a circular connected-(1, 2)-or-(2, 1)-out-of-(m, n):F lattice system.

Combinatorial approaches have also been widely used for evaluating system reliability in the i.i.d. case. El-Sayed (1998) presented the combinatorial approach for evaluating the reliability of a system in the case of $m = 2$. Ishikawa et al. (2016) also provided the combinatorial approach for evaluating the reliability of a system in the case of $m = 2$; it has a simpler form than the equation of El-Sayed (1998). Further, they gave the combinatorial approach for the reliability of a system in the case of $m = 3$.

Most studies focused on the system in the linear and circular cases; however, there are practical systems which cannot be evaluated by existing system models because they have a specific form, e.g., toroidal systems. To evaluate toroidal systems, Nakamura et al. (2016) introduced a toroidal connected-(1, 2)-or-(2, 1)-out-of-(m, n):F lattice system. It is a circular connected-(1, 2)-or-(2, 1)-out-of-(m, n):F lattice system where the components in the 1st circle are connected to those in the m-th circle; thus, this system forms a torus. The components in this system are arranged at the intersections of m circles and n rings, where "circle" and "ring" are shown in Fig. 6. This system fails if and only if at least two consecutive components fail on a circle or a ring.

In Eqs. (21) and (22), let $x_{i0} = x_{in}$ for $i = 1, 2, ..., m$ and $x_{0j} = x_{mj}$ for $j = 1, 2, ..., n$ for the sake of convenience. Let $R^T(m, n; [p_{ij}])$ be the reliability of a toroidal connected-(1, 2)-or-(2, 1)-out-of-(m, n):F lattice system with the reliability p_{ij} of component (i, j) for $i = 1, 2, ..., m$ and $j = 1, 2, ..., n$, and it is defined by

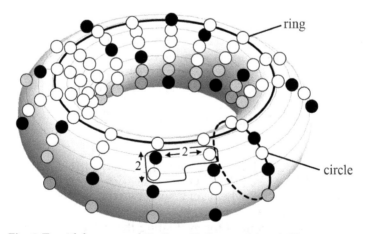

Fig. 6: Toroidal connected-(1, 2)-or-(2, 1)-out-of-(m, n):F lattice system

$$R^T(m, n; [p_{ij}]) = \Pr\left\{\bigcap_{i=1}^{m}\bigcap_{j=1}^{n}\left\{\left\{Z_{ij}^{(1)} = 0\right\}\cap\left\{Z_{ij}^{(2)} = 0\right\}\right\}\right\}. \tag{25}$$

Nakamura et al. (2017(a)) developed an evaluation method for the reliability of a toroidal connected-(1, 2)-or-(2, 1)-out-of-(m, n):F lattice system by using FMCIA. Now, we present the method. The elements of a transition probability matrix are defined as

$$m_{ba}(i) = \prod_{k=1}^{n} p_{ik}^{(1-x_{ik})} q_{ik}^{x_{ik}} \times \phi(\mathbf{x}_i, \mathbf{x}_{i-1}), \tag{26}$$

where $\mathbf{x}_i = (x_{i1}, x_{i2},...x_{in})$ $(i = 1, 2, ..., m)$, $a = 1+\sum_{j-1}^{n}2^{j-1}x_{i-1,j}$, $b = 1+\sum_{j-1}^{n}2^{j-1}x_{i,j}$ and

$$\phi(\mathbf{x}_i, \mathbf{x}_{i-1}) = \prod_{j=1}^{n}[(1 - x_{ij}x_{i-1,j})(1 - x_{ij}x_{i,j-1})(1-x_{i-1,j}x_{i-1,j-1})].$$

The transition probability matrix is given by

$$M(i) = (m_{ba}(i))_{2^n \times 2^n}. \tag{27}$$

Then, the reliability of the toroidal connected-(1, 2)-or-(2, 1)-out-of-(m, n): F lattice system is

$$R^T(m, n; [p_{ij}]) = Tr\left(\prod_{i=1}^{m}M(i)\right), \tag{28}$$

where $Tr(A)$ means a trace of matrix A.

3.3 Connected-$\bigcup_{\theta=1}^{k}X_\theta$-out-of-(*m*, *n*):F lattice systems

In this subsection, we consider connected $\bigcup_{\theta=1}^{k}X_\theta$-out-of-(m, n):F lattice systems. X_θ represents one of any connected failure patterns for $\theta = 1, 2,$ ..., k. For example, X_1 may represent the failure pattern as shown in Fig. 7. A linear connected-X_1-out-of-(7, 7):F system fails when failed components constitute failure pattern X_1, as shown in Fig. 7 (a). However, this system does not fail, even if failed components constitute the failure patterns, which resemble X_1, as shown in Fig. 7 (b). This system fails if and only if there are failed components constituting one of X_θ in the system.

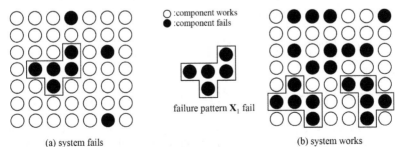

(a) system fails   failure pattern X_1 fail   (b) system works

○ :component works
● :component fails

Fig. 7: Linear connected-X_1-out-of-(7, 7):F system

Yamamoto (1996(a)) considered the limit theorem for a linear or circular connected-$\bigcup_{\theta=1}^{k}X_{\theta}$-out-of-$(m, n)$:F lattice system. For $\varphi = L, C$, let $R^{\varphi}\left(\bigcup_{\theta=1}^{k}X_{\theta}, (m,n); p_n\right)$ be the reliability of a connected-$\bigcup_{\theta=1}^{k}X_{\theta}$-out-of-$(m, n)$:F lattice system with component reliability p_n in the i.i.d. case. Note that the system is linear (circular) if $\varphi = L$ (C), respectively. Let N be the number of failure patterns where failure pattern X_{θ} consists of exactly M components, that is, $\#\{\theta \mid \#X_{\theta} = M\}$ where $M = \min_{1\le\theta\le k}(\#X_{\theta})$. Please note that $\#X_{\theta}$ means the number of components within the failure pattern X_{θ}. Also, let λ be a positive constant value. If $\lim_{n\to\infty} mn(1-p_n)^M = \lambda$, then

$$\lim_{n\to\infty} R^{\varphi}\left(\bigcup_{\theta=1}^{k}X_{\theta}, (m, n); p_n\right) = \exp[-N\lambda], \qquad (29)$$

for $\varphi = L, C$.

Furthermore, Yamamoto (1996(a)) provided the upper and lower bounds for the system reliability by using the following notations.

r_{θ}: number of columns of the smallest rectangle of the rectangles including X_{θ}, as shown in Fig. 8.

s_{θ}: number of rows of the smallest rectangle of the rectangles including X_{θ}, as shown in Fig. 8.

$\gamma_{\theta(i,j)}$: index function, which takes 0, when all components fail in X_{θ} included in the rectangle of size $r_{\theta} \times s_{\theta}$ with component (i, j) as its upper left corner, and takes 1 otherwise.

Let $r_{\max} = \max_{1\le\theta\le k} r_{\theta}$, $s_{\max} = \max_{1\le\theta\le k} s_{\theta}$, $r_{\min} = \min_{1\le\theta\le k} r_{\theta}$ and $s_{\min}= \min_{1\le\theta\le k} s_{\theta}$

$\mathbf{B}_{(i,j)}$: event $\left\{\prod_{\mu=j}^{j+s_{\max}-1}(1-x_{i-1,\mu})\prod_{v=i}^{i+r_{\max}-1}(1-x_{v,j-1}) = 1\right\}$, that is, the event that all the components work, which adjoins upper and left boundaries of the $r_{\theta} \times s_{\theta}$ rectangle with component (i,j) as the upper left corner, as shown in Fig. 8.

By using the above notation, the bounds for the reliability of a linear connected-$\bigcup_{\theta=1}^{k}X_{\theta}$-out-of-$(m, n)$:F lattice system are given by

$$LB = \prod_{i=1}^{m-r_{\min}+1}\prod_{j=1}^{n-s_{\min}+1}\prod_{\theta=1}^{k}[1-\Pr\{\gamma_{\theta(i,j)} = 0\}], \qquad (30)$$

and

$$UB = \prod_{i=1}^{m-r_{\min}+1}\prod_{j=1}^{n-s_{\min}+1}\left[1-\Pr\{\mathbf{B}_{(i,j)}\}\Pr\left\{\bigcup_{\theta=1}^{k}\gamma_{\theta,(i,j)} = 0\right\}\right]. \qquad (31)$$

A connected-(r_1, s_1)-or-(r_2, s_2)-...-(r_k, s_k)-out-of-(m, n):F lattice system fails whenever there exists at least one submatrix of r_i rows and s_i columns for $i = 1, 2, ..., k$, and it is a special case of connected-$\bigcup_{\theta=1}^{k}X_{\theta}$-out-of-$(m, n)$:F

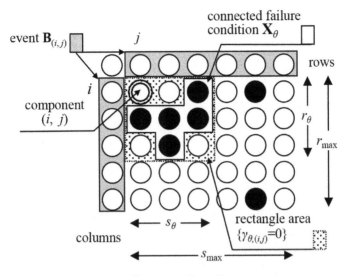

Fig. 8: Notations, $\mathbf{B}_{(i,j)}$, $\mathbf{X}_\theta$, $\gamma_{\theta(i,j)}$

lattice systems. For this system, Yamamoto (1996(b)) developed recursive algorithms for the reliability of the systems in the linear and circular cases.

4. Two-dimensional consecutive-*k*-out-of-*n* related systems and multi-dimensional consecutive-*k*-out-of-*n* systems

In this section, we present the evaluation methods for the reliability of two-dimensional consecutive-*k*-out-of-*n* related systems and multi-dimensional consecutive-*k*-out-of-*n* systems.

4.1 *k*-within-consecutive-(*r*, *s*)-out-of-(*m*, *n*):F system

A linear (circular) *k*-within-consecutive-(*r*, *s*)-out-of-(*m*, *n*):F system consists of *m* × *n* components, arranged like the elements of an *m* × *n* matrix (located on the intersections of *m* circles and *n* rays). It fails if and only if *k* components in an *r* × *s* submatrix fail, as shown in Fig. 9 (a) (Fig. 9 (b)). Random variable Y'_{ij} is redefined as

$$Y'_{ij} = \sum_{a=i-r+1}^{i} \sum_{b=j-s+1}^{j} x_{ab}. \tag{32}$$

Please note that binary variable x_{ij} means the state of component (i, j) ($i = 1, 2, \ldots, m$ and $j = 1, 2, \ldots, n$) as same as Eq. (9).

In other words, Y'_{ij} means the number of failed components in the $r \times s$ matrix with (i, j) at the lower right corner. For $\varphi = L, C$, let $R^\varphi(k, r, s,$

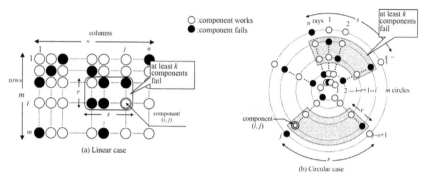

Fig. 9: k-within-consecutive-(r, s)-out-of-(m, n):F system

$m, n;[p_{ij}])$ be the reliability of the k-within-consecutive-(r, s)-out-of-(m, n):F system with reliability p_{ij} of component (i, j), where the system is linear (circular) if $\varphi = L$ (C), respectively. Then,

$$R^L(k, r, s, m, n; [p_{ij}]) = \Pr\left\{\bigcap_{i=r}^{m} \bigcap_{j=s}^{n} \left\{Y_{ij} < k\right\}\right\}, \tag{33}$$

and

$$R^C(k, r, s, m, n; [p_{ij}]) = \Pr\left\{\bigcap_{i=r}^{m} \bigcap_{j=s}^{n+s-1} \left\{Y_{ij} < k\right\}\right\}. \tag{34}$$

If $k = r \times s$, then the k-within-consecutive-(r, s)-out-of-(m, n):F systems are equivalent to a connected-(r, s)-out-of-(m, n):F lattice system.

Lin and Zuo (2000) and Akiba and Yamamoto (2001) proposed the recursive formulas for the reliability of the linear k-within-consecutive-(r, s)-out-of-(m, n):F system independently, and Akiba and Yamamoto (2001) and Yamamoto and Akiba (2004) proposed a recursive algorithm for the reliability of the circular k-within-consecutive-(r, s)-out-of-(m, n):F system. When the system reliabilities are computed by the recursive algorithm of Akiba and Yamamoto (2001), the order of computing time is $\min(O(mk^{r(n-s)}), O(nk^{s(m-r)}))$ and $O(mk^{rn})$ time in the linear and the circular cases, respectively. Chang and Huang (2010) used FMCIA to compute the reliability of the linear and circular k-within-consecutive-(r, s)-out-of-(m, n):F systems.

There are many papers on the bounds for the reliability of the k-within-consecutive-(r, s)-out-of-(m, n):F system. For example, Malinowski and Preuss (1995(a)); Makri and Psillakis (1996, 1997); Akiba et al. (2000); Akiba and Yamamoto (2002) and Yamamoto and Akiba (2005(b)) for a linear system and Makri and Psillakis (1996, 1997); Akiba et al. (2002) for a circular system.

Akiba and Yamamoto (2002) and Akiba et al. (2002) proposed the limit theorems for linear or circular k-within-consecutive-(r, s)-out-of-(m, n):F system in the i.i.d. case. Let λ be a positive constant value. If $\lim_{n \to \infty} mn(1 - p_n)k = \lambda$, then

$$\lim_{n \to \infty} R^L(k,r,s,m,n;p_n) = \exp[-N(k)\lambda], \tag{35}$$

and

$$\lim_{n \to \infty} R^C(k,r,s,m,n;p_n) = \exp[-N(k)\lambda], \tag{36}$$

where

$$N(k) = \binom{rs}{k} - \binom{(r-1)s}{k} - \binom{r(s-1)}{k} + \binom{(r-1)(s-1)}{k}.$$

Akiba and Yamamoto (2002) proposed another limit theorem when n approaches infinity and m is constant. Let λ be a positive constant value and m be a fixed integer. If $\lim_{n \to \infty} n(1-p_n)^k = \lambda$, then

$$\lim_{n \to \infty} R^L(k,r,s,m,n;p_n) = \exp[-(N_r(k) + (m-r)N(k))\lambda], \tag{37}$$

where

$$N_r(k) = \binom{rs}{k} - \binom{r(s-1)}{k}.$$

4.2 Two-dimensional consecutive-*k*-out-of-*n* related systems

Yamamoto and Miyakawa (1996(a)) introduced an adjacent triangle-(m, n):F triangular lattice system. This system consists of $m \times n$ components, which are arranged on an adjacent triangle of a rectangular grid, as shown in Fig. 10. This system fails if and only if at least three components on a triangle fail. Yamamoto et al. (2006) derived the recursive formulas for evaluating the reliability of an adjacent triangle-(m, n):F triangular lattice system.

Zuo et al. (2000) defined a system which involves some failure criteria. A combine k-out-of-mn:F and linear connected-(r, s)-out-of-(m, n):F lattice system has mn components arranged in m rows with n components in each row. This system fails if and only if at least one of the following conditions holds.

(1) At least k components fail in the system.
(2) There exists at least one submatrix (size $r \times s$) of all failed components.

If $k = mn$, it is the same as the linear connected-(r, s)-out-of-(m, n):F lattice system.

A combined k-out-of-mn:F, consecutive-k_c-out-of-n:F and linear connected-(r, s)-out-of-(m, n):F system has mn components arranged in m rows with n components in each row. This system fails if and only if at least one of the following conditions holds.

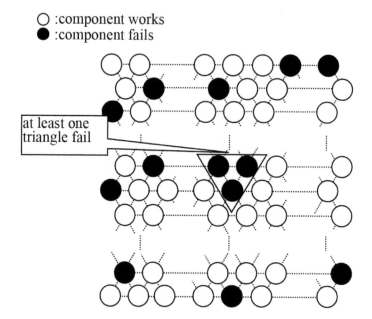

Fig. 10: Adjacent triangle: F triangular lattice system fails

(1) At least k components fail in the system.
(2) At least one row has at least consecutive k_c failed components.
(3) There exists at least one submatrix (size $r \times s$) of all failed components.

Yuge et al. (2000) proposed a conditioned linear connected-(k, k)-out-of-(n, n):F lattice system. This system also has n^2 components arranged in n rows with n components in each row, and fails if and only if

(1) there are M failed components in the system ($M \geq k^2$), or
(2) there are more than or equal to M_0 failed components in an arbitrary submatrix of size $k \times k$ ($M_0 \leq k^2$).

This system may be said to be a combined system of M_0-within-consecutive-(k, k)-out-of-(n, n):F system and M-out-of-n^2: F system. Yuge et al. (2000) discussed availability and MTBF of this system with maintenance. In particular, when $M_0 = k^2$, $M \leq k^2$, that is, for a combined M-out-of-n^2:F and linear connected-(k, k)-out-of-(n, n):F system, they proposed a closed formula for availability and MTBF when $k = 2, 3, 4$. Yuge et al. (2003) derived closed formulas for the reliability of the combined $2k^2$-out-of-n^2:F and linear connected-(k, k)-out-of-(n, n):F system in the i.i.d. case when $k = 2, 3, 4$. Habib et al. (2010) extended the conditioned linear connected-(k, k)-out-of-(n, n):F lattice system to a conditioned linear connected-(r, s)-out-of-(m, n):F lattice system.

4.3 Multi-dimensional consecutive-*k*-out-of-*n* systems

In this subsection, we review the studies on the evaluation methods for the reliability of multi-dimensional consecutive-*k*-out-of-*n* systems. A three-dimensional consecutive-*k*-out-of-*n* system consists of $n_1 \times n_2 \times n_3$ components, which are arranged like a (n_1, n_2, n_3) rectangular solid. Salvia and Lasher (1990) showed promising applications of such multi-dimensional models; e.g., the presence of a disease is diagnosed by using multiple X-ray or MRI, etc. This application gave the probability that an individual cell (or other small portions of the X-ray) is healthy or sick. Unless disease cells are aggregated into a sufficiently large pattern (say a $k \times k$ square), the radiologist might not detect their presence. In medical diagnostics, it may be more appropriate to consider a three-dimensional grid in order to calculate the detection probability of patterns in a three-dimensional space. In another example, a three-dimensional consecutive-*k*-out-of-*n* system can be applied for the mathematical model of a three-dimensional flash memory cell failure model, and hypercube topology of the connection network, and so on (Akiba and Yamamoto 2004 etc.).

Because of the complexity of the system, several researchers provided the upper and lower bounds and the limit theorems for the system reliability or the exact reliability evaluation methods for special cases of such systems.

Boushaba and Ghoraf (2002) proposed upper and lower bounds and a limit theorem for this system. Godbole et al. (1998) proposed an upper bound for the reliability of a *d*-dimensional consecutive-*k*-out-of-*n*:F system. A three-dimensional *k*-within-consecutive-(r_1, r_2, r_3)-out-of-(n_1, n_2, n_3):F system is a three-dimensional version of the linear *k*-within-consecutive-(r, s)-out-of-(m, n):F system, and it fails if and only if there is an (r_1, r_2, r_3) rectangular solid in which *k* or more components fail, as shown in Fig. 11. Akiba and Yamamoto (2004) provided the upper and lower bounds for this system.

Gharib et al. (2010, 2011) proposed the exact evaluation methods for the reliability of the special cases of three-dimensional consecutive-*k*-out-of-*n* systems. Cowell (2015) derived general closed-form exact formulae for the reliability of a *d*-dimensional consecutive-*k*-out-of-*n*:F system based on the inclusion-exclusion principle. However, we feel four or more dimensional systems are considered under the mathematical interest, but we think these systems may be useful in statistical models for uniformity test between many variables. We hope that the future work on these systems will give us useful results.

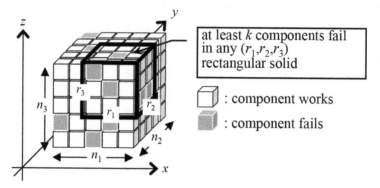

Fig. 11: Example of failure of three-dimensional k-within-consecutive-(r_1, r_2, r_3)-out-of-(n_1, n_2, n_3):F system

5. Conclusion and future works

In this chapter, we considered to the consecutive-k systems. We showed definitions and the evaluation methods for the consecutive-k systems, that is, consecutive-k-out-of-n system, multi-dimensional consecutive-k-out-of-n systems and their related systems. In recent years, other evaluation methods for the consecutive-k systems were proposed, for example, optimal component arrangement problem, component importance problem, system signature problem, etc. This chapter does not refer to the methods of obtaining solutions for these problems. We need more surveys for evaluation methods for the consecutive-k and its related systems, for example, insight on these models, particularly their mathematical relationships for the consecutive-k systems. These are important future work.

REFERENCES

Akiba, T. and H. Yamamoto. 2001. Reliability of a 2-dimensional k-within-consecutive-$r \times s$-out-of-$m \times n$:F system. Nav. Res. Logist. 48(7): 625–637.

Akiba, T. and H. Yamamoto. 2002. Evaluation for the reliability of a large 2-dimensional rectangular k-within-consecutive-(r, s)-out-of-(m, n):F system. J. Japan Ind. Manag. Assoc. 53(3): 208–219. (In Japanese)

Akiba, T. and H. Yamamoto. 2004. Upper and lower bounds for 3-dimensional k-within-consecutive-(r_1, r_2, r_3)-out-of-(n_1, n_2, n_3):F system. Proc. 2004 Asian Int. Workshop on Adv. Reliab. Model. 9–16.

Akiba, T., H. Yamamoto and W. Saitou. 2000. Upper and lower bounds for reliability of 2-dimensional-k-within-consecutive-(r, s)-out-of-(m, n):F system. Reliab. Eng. Assoc. Japan 22: 99–106. (In Japanese)

Akiba, T., H. Yamamoto and H. Tsubone. 2002. Approximate values of reliability of the 2-dimensional cylindrical *k*-within-consecutive-(*r*, *s*)-out-of-(*m*, *n*):F system. Proc. Fourth Asia-Pacific Conf. Ind. Eng. Manag. Syst.

Beiu, V. and L. Daus. 2015. Reliability bounds for two dimensional consecutive systems. Nano Commun. Netw. 6(3): 145–152. Elsevier Ltd.

Boehme, T.K., A. Kossow and W. Preuss. 1992. A generalization of consecutive-*k*-out-of-*n*:F system. IEEE Trans. Reliab. 41(3): 451–457.

Boutsikas, M.V. and M.V. Koutras. 2000. Generalized reliability bounds for coherent structures. J. Appl. Probab. 37(3): 778–794.

Chang, G.J., L. Cui and F.K. Hwang. 2000. Reliabilities of consecutive-*k* systems (Network Theory and Applications, Volume 4), Kluwer Academic Publishers, Dordrecht.

Chang, Y. and T. Huang. 2010. Reliability of a 2-dimensional *k*-within-consecutive-*r* × *s*-out-of-*m* × *n*:F system using finite Markov chains. IEEE Trans. Syst. Reliab. 59(4): 725–733.

Chao, M., T.J.C. Fu and M.V. Koutras. 1995. Survey of reliability studies of consecutive-*k*-out-of-*n*:F related systems. IEEE Trans. Reliab. 44(1): 120-127.

Cowell, S. 2015. A formula for the reliability of a *d*-dimensional consecutive-*k*-out-of-*n*:F system. Int. J. Comb. 2015: 1–6.

Cui, L., W. Kuo, J. Li and M. Xie. 2006. On the dual reliability systems of (*n*, *k*, *j*) and < *n*, *k*, *f* >. Stat. Probab. Lett. 76: 1081–1088.

Cui, L., Y. Xu and X. Zhao. 2010. Developments and applications of the finite Markov chain imbedding approach in reliability. IEEE Trans. Reliab. 59(4): 685–690.

El-Sayed, E.M. 1998. Algorithm for reliability of (1,2) or (2,1)-out-of (*n*, 2):F systems. J. Egypt. Math. Soc. 6(2): 169–173.

Eryilmaz, S. 2010. Review of recent advances in reliability of consecutive *k*-out-of-*n* and related systems. Proc. Inst. Mech. Eng. Part O J. Risk Reliab. 224(3): 225–237.

Fu, J.C. 1986. Reliability of consecutive-*k*-out-of-*n*:F systems with (*k* − 1)-step Markov dependence. IEEE Trans. Reliab. 35(5): 602–606.

Fu, J.C. and B. Hu. 1987. On reliability of a large consecutive-*k*-out-of-*n*:F system with (*k* − 1)-step Markov dependence. IEEE Trans. Reliab. 36(1): 75–77.

Fu, J.C. and M.V. Koutras. 1994. Poisson approximation for 2-dimensional patterns. Ann. Inst. Stat. Math. 46 (1): 179–192.

Gharib, M., E.M. El-Sayed and I. Nashwan. 2010. Reliability of simple 3-dimensional consecutive *k*-out-of-*n*:F systems. J. Math. Stat. 6(3): 261–264.

Gharib, M., E.M. El-Sayed and I. Nashwan. 2011. Reliability of connected (1,1,2) or (1,2,1) or (2,1,1)-out-of-(*n*,2,2):F lattice systems. J. Adv. Res. Stat. Probab. 3(1): 47–56.

Godbole, A.P., L.K. Potter and J.K. Sklar. 1998. Improved upper bounds for the reliability of *d*-dimensional consecutive-*k*-out-of-*n*:F systems. Nav. Res. Logist. 45(2): 219-230.

Habib, A.S., T. Yuge, R.O. Al-Seedy and S.I. Ammar. 2010. Reliability of a consecutive-(*r*, *s*)-out-of-(*m*, *n*):F lattice system with conditions on the number of failed components in the system. Appl. Math. Model. 34(3): 531–538. Elsevier Inc.

Hwang, F.K. 1982. Fast solutions for consecutive-k-out-of-n:F systems. IEEE Trans. Reliab. 31(5): 447–448.

Ishikawa, T., T. Shinzato, T. Nakamura and H. Yamamoto. 2016. Expansion on Ratio q/p for reliability function of linear connected-(1,2)-or-(2,1)-out-of-(m, n):F lattice system. Proc. 7th Asia-Pacific Int. Symp. Adv. Reliab. Maint. Model. 162–169.

Kontoleon, J.M. 1980. Reliability determination of a r-successive-out-of-n:F system, IEEE Trans. Reliab. 29(5): 29–47.

Koutras, M.V. 1996. On a Markov chain approach for the study of reliability structures. J. Appl. Probab. 33(2): 357–367.

Koutras, M.V., G.K. Papadopoulos and S.G. Papastavridis. 1997. A reliability bound for 2-dimensional consecutive-k-out-of-n:F systems. Nonlinear Anal. Theory, Methods Appl. 30(6): 3345–3348.

Koutras, M.V. and S.G. Papastavridis. 1993. Application of the Stein-Chen method for bounds and limit theorems in the reliability of coherent structure. Nav. Res. Logist. 40(5): 617–631.

Ksir, B. 2000. A weibull limit law for a "2-dimensional consecutive-k-out-of-n":F system. Proc. Second Int. Conf. Math. Methods Reliab. Bordeaux, Fr. Bordeaux, Univers. Victor-Segalen. 651–654.

Kuo, W. and M.J. Zuo. 2002. Optimal reliability modeling: principles and applications. John Wiley & Sons Inc., Hoboken, New Jersey.

Lin, C., L. Cui, D.W. Coit and M. Lv. 2016. Reliability modeling on consecutive-k_r-out-of-n_r:F linear zigzag structure and circular polygon structure. IEEE Trans. Reliab. 65(3): 1509–1521.

Lin, D. and M.J. Zuo. 2000. Reliability evaluation of a linear k-within-(r, s)-out-of-(m, n):F lattice system. Probab. Eng. Informational Sci. 14(4): 435–443.

Makri, F.S. and Z.M. Psillakis. 1996. Bounds for reliability of k-within two-dimensional consecutive-r-out-of-n failure systems. Microelectro. Reliab. 36(3): 341–345.

Makri, F.S. and Z.M. Psillakis. 1997. Bounds for reliability of k-within connected-(r, s)-out-of-(m, n) failure systems. Microelectro. Reliab. 37(8): 1217–1224.

Malinowski, J. and W. Preuss. 1995(a). On the reliability of generalized consecutive systems – A survey. Int. J. Reliab. Qual. Saf. Eng. 2(2): 187–201.

Malinowski, J. and W. Preuss. 1995(b). A recursive algorithm evaluating the exact reliability of a consecutive k-within-m-out-of-n:F system. Microelectro. Reliab. 35(12): 1461–1465.

Malinowski, J. and W. Preuss. 1996(a). Lower & upper bounds for the reliability of consecutive-(r, s)-out-of-(m, n):F lattice systems. IEEE Trans. Reliab. 45(1): 156–160.

Malinowski, J. and W. Preuss. 1996(b). A recursive algorithm evaluating the exact reliability of a circular consecutive-k-within-m-out-of-n:F system. Microelectro. Reliab. 36(10): 1389–1394.

Nakamura, T., H. Yamamoto, T. Shinzato, T. Akiba and X. Xiao. 2016. Reliability of a toroidal connected-(1,2)-or-(2,1)-out-of-(m, n):F lattice system. Proc. 7th Asia-Pacific Int. Symp. Adv. Reliab. Maint. Model. 399–406.

Nakamura, T., H. Yamamoto, T. Shinzato, X. Xiao and T. Akiba. 2017(a). Proposal of calculation method for reliability of toroidal connected-(1,2)-or-(2,1)-out-of-(m, n):F lattice system with Markov chain. pp. 139–153. *In*: S. Nakamura, C.H. Qian and T. Nakagawa (eds.). Reliab. Model. with Comput. Maint. Appl. World Scientific. Singapore.

Nakamura, T., H. Yamamoto, T. Shinzato, X. Xiao and T. Akiba. 2017(b). Reliability of a circular connected-(1,2)-or-(2,1)-out-of-(m, n):F lattice system with identical components. IEICE Trans. Fundam. Electron. Commun. Comput. Sci. E100A (4): 1029–1036.

Nakamura, T., H. Yamamoto and X. Xiao. 2018. Fast calculation methods for reliability of connected-(r, s)-out-of-(m, n):F lattice system in special cases. Int. J. Math. Eng. Manag. Sci. 3(2): 113–122.

Noguchi, K., M. Sasaki, S. Yanagi and T. Yuge. 1996. System reliability of a sensor system considering the detectable area. IEICE Trans. Fundam. Electron. Commun. Comput. Sci. J79–A (4): 1444–1453. (in Japanese)

Salvia, A.A. and W.C. Lasher. 1990. 2-dimensional consecutive-k-out-of-n:F models. IEEE Trans. Reliab. 39(3): 382–385.

Triantafyllou, I.S. 2015. Consecutive-type reliability systems: an overview and some applications. J. Qual. Reliab. Eng. 2015: 20 pages.

Yamamoto, H. 1996(a). Reliability of connected-X-out-of-(m, n):F lattice system. Proc. Inst. Stat. Math. 44(2): 201–210. (in Japanese)

Yamamoto, H. 1996(b). Reliability of a connected-(r_1, s_1) – or-(r_2, s_2)-or-... -or-(r_k, s_k)-out-of-(m, n):F lattice system. Microelectro. Reliab. 36(2): 151–168.

Yamamoto, H. and T. Akiba. 2005(a). A recursive algorithm for the reliability of a circular connected-(r, s)-out-of-(m, n):F lattice system. Comput. Ind. Eng. 49(1): 21–34.

Yamamoto, H. and T. Akiba. 2005(b). Evaluating methods for the reliability of a large 2-dimensional rectangular k-within-consecutive-(r, s)-out-of-(m, n):F system. Nav. Res. Logist. 52(3): 243–252.

Yamamoto, H., T. Akiba, H. Nagatsuka and Y. Moriyama. 2008. Recursive algorithm for the reliability of a connected-(1,2)-or-(2,1)-out-of-(m, n):F lattice system. Eur. J. Oper. Res. 188(3): 854–864.

Yamamoto, H., T. Akiba and Y. Wakagi. 2006. Efficient algorithm for the reliability of an adjacent triangle:F triangular lattice system. J. Reliab. Eng. Assoc. Japan 28(1): 63–76.

Yamamoto, H. and M. Miyakawa. 1995. Reliability of linear connected-(r, s)-out-of-(m, n):F lattice system. IEEE Trans. Reliab. 44(2): 333–336.

Yamamoto, H. and M. Miyakawa. 1996(a). The reliability of the adjacent triangle:F triangular lattice system. J. Reliab. Eng. Assoc. Japan 18(7): 97–105. (in Japanese)

Yamamoto, H. and M. Miyakawa. 1996(b). Reliability of circular connected-(r, s)-out-of-(m, n):F lattice system. J. Oper. Res. Soc. of Japan 39(3): 389–406.

Yuge, T., M. Dehare and S. Yanagi. 2000. Reliability and availability of a repairable lattice system. IEICE Trans. Fundam. Electron. Commun. Comput. Sci. E83–A (5): 782–787.

Yuge, T., M. Dehare and S. Yanagi. 2003. Reliability of a 2-dimensional consecutive-k-out-of-n:F system with a restriction in the number of failed components. IEICE Trans. Fundam. Electron. Commun. Comput. Sci. E86–A (6): 1535–1540.

Zhao, X. and L. Cui. 2009. On the accelerated scan finite Markov chain. IEEE Trans. Reliab. 58(2): 383–388.

Zhao, X., L. Cui, W. Zhao and F. Liu. 2011. Exact reliability of a linear connected-(r, s)-out-of-(m, n): F system. IEEE Trans. Reliab. 60(3): 689–698.

Zuo, M.J. 1993. Reliability and design of 2-dimensional consecutive-k-out-of-n:F systems. IEEE Trans. Reliab. 42(3): 488–496.

Zuo, M.J., D. Lin and Y. Wu. 2000. Reliability evaluation of combined k-out-of-n:F, consecutive-k_c-out-of-n:F and linear connected-(r, s)-out-of-(m, n):F system structures. IEEE Trans. Reliab. 49(1): 99–104.

Reliability Properties of *K*-out-of-*N*:*G* Systems

K. Ito[1*]**and T. Nakagawa**[2]

[1] Department of Social Management Engineering, Tottori University,
   Minami 4-101, Koyama, Tottori-shi, 680–8552, Japan
[2] Department of Business Administration, Aichi Institute of Technology,
   1247 Yachigusa, Yagusa-cho, Toyota, Aichi, 470–0392, Japan

1. Introduction

A highly reliable system can be realized by employing redundancy. A standard parallel system, which consists of N identical subsystem, is the most typical redundant model. It was shown illustratively that the redundant system can be utilized for a specified mean time by either changing the replacement time or increasing the number of subsystems (Barlow and Proschan 1965). In addition, qualitative disparities between system reliability and subsystem (Barlow and Proschan 1965), and reliabilities of many redundant systems were discussed (Ushakov 1994). A variety of redundant systems with multiple failure modes (Pham 2003) and dependent failures of units (Blokus 2006) were considered. Furthermore, some optimization methods of redundancy allocation for series-parallel systems were surveyed (Zia and Coit 2010). The variety of redundant systems affects maintenance policies. Lately, a new classification of three preventive maintenance models was suggested (Wu and Zuo 2010) and optimal maintenace policies for 2-phase systems with worn states were investigated (MacPherson and Glazebrook 2011).

A K-out-of-N ($1 \le K \le N$) system can operate if and only if at least K units of the total N units are operable (Barlow and Proschan 1965). The same is widely used in practical systems, such as data transmission, redundant networks and redundant copies because such systems can

*Corresponding author: itokodo@tottori-u.ac.jp

realize high reliability with moderately reliable components (Nakagawa 2008). Their reliability characteristics were also investigated (Linton and Saw 1974, Nakagawa 1985). The managers of mass transits and computer systems have to put N units on line with the assurance that at least K units will be available to complete the mission, to that end, their solutions for computing availabilities of K-out-of-N:G systems were provided (Kenyon and Newell 1983). A K-out-of-N:G system is also used as a self-checking checker for error detecting codes (Lala 1985). Studies of multi-state and consecutive K-out-of-N:G systems were surveyed (Chang, Cui and Hwang 2000, Kuo and Zuo 2003).

In this chapter, the conventional K-out-of-N:G system, in which both K and N are constant, and the random K-out-of-N:G system, in which K changes stochastically, are considered.

In Section 2, the basic properties of reliability and mean time to failure (MTTF) are discussed. K-out-of-N:G system and 1-unit system, whose reliabilities are equivalent, are presented when K and N are varied. Minimum Ns, which exceed specific MTTFs, are depicted when K and MTTF are alternated.

In Section 3, a random K-out-of-N system is introduced and its reliability and MTTF are assigned. Because K changes statistically with the characteristics that may differ in system structures and conditions, K is assumed to be a random variable with a probability function $p_k \equiv \Pr\{K = k\}$ ($k = 0, 1, 2, \cdots , N$). As a special case of a random K-out-of-N system, random K-out-of-4 systems are proposed in Section 4. Reliabilities and MTTFs of these two cases are mentioned. In Section 5, hierarchical structured systems are considered. A "hierarchical" system is a system composed of sub-systems, of which the structures are the same, and the system totalising the sub-systems is also the same structure. For example, the upper stage system is 2-out-of-4 when the lower stage one is 2-out-of-4. In Section 6, practical applications of the random K-out-of-N:G system are argued as concluding remarks of this chapter (Ito, Zhao and Nakagawa 2017). Such results would be greatly useful for designing K-out-of-N:G systems and making their appropriate maintenances practically in actual fields.

2. Reliability of *K*-out-of-*N*:G system

When each unit has an identical failure distribution $F(t)$ with a finite mean $\mu \equiv \int_0^\infty \overline{F}(t)\,dt < \infty$ and a density function $f(t) \equiv dF(t)/dt$ where $\overline{F}(t) \equiv 1 - F(t)$, the reliability of the system at time t is (Barlow and Proschan 1965, Nakagawa 2005)

$$R\left(\overline{F}(t)\right) = \sum_{j=0}^{N-K} \binom{N}{j} \left[F(t)\right]^j \left[\overline{F}(t)\right]^{N-j}$$

$$= \sum_{j=K}^{N} \binom{N}{j} \left[\overline{F}(t) \right]^{j} \left[F(t) \right]^{N-j} \tag{1}$$

It can easily be shown that $R(\overline{F}(t))$ decreases with K and increases with N because

$$\sum_{j=K}^{N+1} \binom{N+1}{j} \left[\overline{F}(t) \right]^{j} \left[F(t) \right]^{N+1-j} - \sum_{j=K}^{N} \binom{N}{j} \left[\overline{F}(t) \right]^{j} \left[F(t) \right]^{N-j}$$

$$= \sum_{j=K}^{N+1} \left[\binom{N}{j} + \binom{N}{j-1} \right] \left[\overline{F}(t) \right]^{j} \left[F(t) \right]^{N+1-j} - \sum_{j=K}^{N} \binom{N}{j} \left[\overline{F}(t) \right]^{j} \left[F(t) \right]^{N-j}$$

$$= \binom{N}{K-1} \left[\overline{F}(t) \right]^{K} \left[F(t) \right]^{N-K+j} \geq 0$$

where $\binom{N}{j} \equiv 0$ for $j > N$.

We compare the reliability of K-out-of-N:G system with 1-unit system ($N = 2, 3, ...$) when each unit has an identical reliability $\overline{F}(t) \equiv p \, (0 < p < 1) [1, 2]$: From the reliability in (1),

$$R(p) = \sum_{j=0}^{N-K} \binom{N}{j} (1-p)^{j} \, p^{N-j} \quad (K = 1, 2, ..., N) \tag{2}$$

In particular, when $K = 1$, i.e., the system is a parallel one, the reliability is

$$R(p) = 1 - (1-p)^{N} > p$$

This shows that the reliability of a parallel system is greater than that of 1-unit system. When $K = N$, i.e., the system is a series one, the reliability is

$$R(p) = p^{N} < p.$$

This shows that the reliability of a series system is smaller than that of 1-unit system. These results have been well-known in reliability theory.

We are concerned with values of p such that the reliability of K-out-of-N:G systems is greater than that of 1-unit system. First, we compare 2-out-of-3 with 1-unit system: The reliability of 2-out-of-3 system is

$$R(p) = p^{3} + 3(1-p)p^{2}.$$

Setting

$$p^{3} + 3(1-p)p^{2} = p,$$

we have $p = 0.5$. That is, the reliability of 2-out-of-3 system is greater than that of 1-unit system when $p > 0.5$, both reliabilities are equal when $p = 0.5$, and the reliability of 2-out-of-3 is smaller than that of 1-unit system when $p < 0.5$.

Similarly, when $N = 4$, for 2-out-of-4 system, setting

$$p^4 + 4(1-p)p^3 + 6(1-p)^2p^2 = p,$$

i.e.,

$$3p^2 - 5p + 1 = 0,$$

we have

$$p = \frac{5 - \sqrt{13}}{6} \approx 0.2324 \cdot$$

For 3-out-of-4 system, setting

$$p^4 + 4(1-p)p^3 = p,$$

i.e.,

$$3p^2 - p - 1 = 0,$$

we have

$$p = \frac{1 + \sqrt{13}}{6} \approx 0.7676$$

Table 1 presents numerical values p such that the reliability of K out-of-N:G system is equal to that of 1-unit system. This indicates that these values p increase strictly with K from 0 to 1 for N and decrease with N from 1 for $K \geq 2$. This also indicates that such values of N-out-of $2N - 2$, N-out-of-$2N - 1$, and N-out-of-$2N$ would decrease, be constant, and increase with N, respectively. For example, for N-out-of-$2N - 1$, the reliability in (1) when $p = 1/2$ is

$$\sum_{j=0}^{N-1} \binom{2N-1}{j} \left(\frac{1}{2}\right)^j \left(\frac{1}{2}\right)^{2N-1-j} = \left(\frac{1}{2}\right)^{2N-1} \sum_{j=0}^{N-1} \binom{2N-1}{j}$$

$$= \left(\frac{1}{2}\right)^{2N-1} \sum_{j=N}^{2N-1} \binom{2N-1}{j},$$

and

$$\sum_{j=0}^{2N-1} \binom{2N-1}{j} = 2^{2N-1}$$

Thus, for any $N \geq 1$, the reliability is

$$\sum_{j=0}^{N-1} \binom{2N-1}{j} \left(\frac{1}{2}\right)^j \left(\frac{1}{2}\right)^{2N-1-j} = \frac{1}{2} \cdot$$

Thus, the reliability of N-out-of-$2N - 1$ system is greater than that of 1-unit system for $1/2 < p < 1$. By exploring the monotonic reliability properties of N-out-of-$2N - 1$ system, the condition that its reliability is greater than that of 1-unit system is clarified.

In addition, we obtain the number N, which satisfies

$$\sum_{j=0}^{N-K} \binom{N}{j} (1-p)^j \, p^{N-j} \geq p_0, \tag{3}$$

for a specified K and p previously, where p_0 is defined as a required reliability of K-out-of-N system and is assigned for evaluation. Table 2 presents N for K and p_0 when $p = 0.9$. Clearly, number N increases with K and p. For example, when $p_0 = 0.99$ and $K = 5$, we have to prepare 8 units in order to hold that the system reliability is greater than 0.99.

Table 1: Values of p where both reliabilities of K-out-of-N and 1-unit systems are the same

K	N								
	2	3	4	5	6	7	8	9	10
1	0	0	0	0	0	0	0	0	0
2	1	0.5000	0.2324	0.1311	0.0836	0.0579	0.0423	0.0323	0.0255
3	—	1	0.7676	0.5000	0.3471	0.2559	0.1974	0.1577	0.1294
4	—	—	1	0.8689	0.6529	0.5000	0.3955	0.3219	0.2683
5	—	—	—	1	0.9164	0.7441	0.6045	0.5000	0.4214
6	—	—	—	—	1	0.9421	0.8026	0.6781	0.5786
7	—	—	—	—	—	1	0.9577	0.8423	0.7317
8	—	—	—	—	—	—	1	0.9677	0.8706
9	—	—	—	—	—	—	—	1	0.9745
10	—	—	—	—	—	—	—	—	1

Table 2: Number N to satisfy (3) when $p = 0.9$.

K	p_0					
	0.5	0.6	0.7	0.8	0.9	0.99
1	1	1	1	1	1	2
2	2	2	2	2	3	4
3	3	3	3	4	4	5
4	4	4	5	5	5	7
5	5	6	6	6	7	8
6	6	7	7	7	8	9
7	8	8	8	8	9	11
8	9	9	9	10	10	12
9	10	10	10	11	11	13
10	11	11	12	12	13	14

The MTTF is, from (1),

$$\mu_{N,K} \equiv \int_0^\infty R\left(\overline{F}(t)\right) dt$$

$$= \sum_{j=0}^{N-K} \binom{N}{j} \int_0^\infty \left[F(t) \right]^j \left[\bar{F}(t) \right]^{N-j} dt$$

$$= \sum_{j=K}^{N} \binom{N}{j} \int_0^\infty \left[\bar{F}(t) \right]^j \left[F(t) \right]^{N-j} dt \tag{4}$$

which decreases with K from $\int_0^\infty \left\{ 1 - \left[F(t) \right]^N \right\} dt$ to $\int_0^\infty \left[\bar{F}(t) \right]^N dt$.

In particular, when $F(t) = 1 - \exp(-\lambda t)$,

$$\mu_{N,K} = \frac{1}{\lambda} \sum_{j=K}^{N} \frac{1}{j} \tag{5}$$

and is approximately given by

$$\mu_{N,K} = \frac{1}{\lambda} \left[\ln N - \ln (K-1) \right] = \frac{1}{\lambda} \ln \frac{N}{K-1}, \tag{6}$$

for large N, which is motivated by Nakagawa and Yun (2011). In case of $K = 1$, $\mu_{N,K} = (1/\lambda)(\ln N + \gamma)$ (Nakagawa and Zhao 2005), where $\gamma \equiv 0.5772156649...$ (Havil 2003) is Euler's constant. Clearly, MTTF $\mu_{N,K}$ decreases with K to $1/(N\lambda)$ and increases with N to ∞ when $F(t)$ is an exponential distribution.

Futhermore, we obtain the minimum number N such that MTTF in (4) is greater than that of 1-unit system, i.e.,

$$\sum_{j=K}^{N} \frac{1}{j} \geq 1. \tag{7}$$

Table 3: Number N which satisfies (7)

K	1	2	3	4	5	6	7	8	9	10
N	1	4	7	10	12	15	18	20	23	26

Table 3 presents the minimum number N which satisfies (7) for K. It is of great interest from (6) that $N \approx (K-1)e + 1$.

Similarly, we obtain the minimum number N, which satisfies

$$\frac{1}{\lambda} \sum_{j=K}^{N} \frac{1}{j} \geq \mu_0 \tag{8}$$

for a specified K and μ_0. Table 4 presents the minimum number N, which satisfies (8) for K and μ_0 when $1/\lambda = 1$. The value of N increases greatly with K and μ_0. This means that when K and MTTF are large, we have to hold systems with numerous units. These values in Tables 1, 2, 3, and 4 would be useful for constructing K-out-of-N:G systems practically in actual fields.

3. Random *K*-out-of-*N*:G system

We propose *K*-out-of-*N*:G systems when *K* is random, and their reliabilities and MTTFs.

Suppose that K is a random variable with a probability function $p_k \equiv \Pr\{K = k\}$ ($k = 0, 1, 2, \dots, N$) for a specified N ($N = 1, 2, \dots$) where $p_0 = 0$, irrespective of the number of failed units. We denote that the distribution of K is $P_k \equiv \Pr\{K \le k\} = \sum_{j=0}^{k} p_j (k = 0, 1, \dots, N)$, where $P_0 \equiv 0$, $P_N = 1$ and P_k increases with k from 0 to 1. Then, the reliability of the system at time t is, from (1),

$$R\left(\bar{F}(t)\right) = \sum_{k=0}^{N} p_k \sum_{j=k}^{N} \binom{N}{j} \left[\bar{F}(t)\right]^j \left[F(t)\right]^{N-j}$$

$$= \sum_{j=0}^{N} \binom{N}{j} \left[\bar{F}(t)\right]^j \left[F(t)\right]^{N-j} \sum_{k=0}^{j} p_k$$

$$= \sum_{k=1}^{N} P_k \binom{N}{k} \left[\bar{F}(t)\right]^k \left[F(t)\right]^{N-k} \tag{9}$$

This means that the reliability of the system at time t is

$$\sum_{k=1}^{N} \Pr\{K \ge k\} \Pr \{k \text{ units are operative at time } t\},$$

which would be very useful for computing the reliability of the system with a random number of K. Thus, MTTF is, from (4),

$$\mu_{N,P} = \sum_{k=1}^{N} P_k \binom{N}{k} \int_0^\infty \left[\bar{F}(t)\right]^k \left[F(t)\right]^{N-k} dt \tag{10}$$

Table 4: Number N to satisfy (8) when $\lambda = 1$

K	MTTF μ_0						
	0.5	0.8	1.0	2.0	3.0	4.0	5.0
1	1	1	1	4	11	31	83
2	2	3	4	11	31	83	227
3	4	6	7	19	51	137	373
4	6	8	10	26	71	192	521
5	7	10	12	33	91	246	669
6	9	12	15	41	111	301	817
7	11	14	18	48	131	355	966
8	12	17	20	55	151	410	1114
9	14	19	23	63	171	464	1262
10	16	21	26	70	191	519	1411

In particular, when

$$P_k = \begin{cases} 0 & k < K - 1 \\ 1 & k \geq K \end{cases}$$

the system corresponds to K-out-of-N:G system, and when $P_k = k/N$ $(k = 0,$ $1, 2, ..., N)$, $R(\overline{F}(t)) = \overline{F}(t)$, $\mu_{N,P} = \int_0^\infty \overline{F}(t)\,dt$ and the system corresponds to 1-unit system stochastically. Furthermore, when $F(t) = 1 - \exp(-\lambda t)$,

4. Examples of random *K*-out-of-4 system

A special case of systems with random K is introduced to a random K-out-of-4 system.

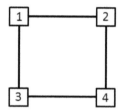

Fig. 1: Random K-out-of-4 system.

4.1 Example 1

As one of practical examples, we consider the following model: The surface of an aluminum-skin airplane is composed of sheet metals which may be 3.8 meters long and 1.5 meters wide. Overlapping portions of sheets are called "lap joints" or "lap splices" and are typically joined together with three rows of hundreds of rivets. Lap joints are another practical example of K-out-of-N:G system. When rivets fail and are lost, an aperture appears because the pressurized inner air of airframe bursts out and deforms the sheet metal. The size of the aperture depends on the number of failed adjacent rivets. When two discontinuous rivets are lost, the aperture size is just twice the size of a one rivet loss aperture. When two contiguous rivets are lost, the aperture size is larger than twice the size of a one rivet loss aperture because the sheet metal deformation in the case of two contiguous rivets being lost is greater than that of two discontinuous rivets being lost. When the bore aperture size exceeds a prespecified size, the airframe cannot operate and has to be repaired. Figure 1 is a random K-out-of-4 system and a simplified example of the airframe. When discontinuous units (unit 1 and 4, 2 and 3) fail, the system can operate; however, it cannot operate when contiguous units (unit 1 and 2, 2 and 4, 1 and 3, 3 and 4) fail. From this model, we consider the simple system which consists of four units, i.e., 1, 2, 3, and 4 in Fig. 1, where when two units have failed, if units 1, 2 or 3, 4 are operative, the system is operative.

When q_i denotes unreliability of units i ($i = 1, 2, 3, 4$) and $1 - q_i$ denotes reliability of units i, the system reliability is

$$R = (1 - q_1)(1 - q_2)(1 - q_3)(1 - q_4)$$

$$+ q_1(1 - q_2)(1 - q_3)(1 - q_4) + (1 - q_1)q_2(1 - q_3)(1 - q_4)$$

$$+ (1 - q_1)(1 - q_2)q_3(1 - q_4) + (1 - q_1)(1 - q_2)(1 - q_3)q_4$$

$$+ q_1(1 - q_2)(1 - q_3)q_4 + (1 - q_1)q_2q_3(1 - q_4),$$

and the unreliability is

$$1 - R = q_1q_2q_3q_4$$

$$+ (1 - q_1)q_2q_3q_4 + q_1(1 - q_2)q_3q_4 + q_1q_2(1 - q_3)q_4$$

$$+ q_1q_2q_3(1 - q_4) + q_1 q_2(1 - q_3)(1 - q_4) + (1 - q_1) q_2 (1 - q_3)$$

$$q_4 + (1 - q_1)(1 - q_2)q_3q_4 + q_1 (1 - q_2)q_3 (1 - q_4).$$

When $q_i = q$,

$$R = (1 - q)^4 + 4q(1 - q)^3 + 2q^2(1 - q)^2 = (1 - q)^2(1+2q - q^2),$$

$$1 - R = q^4 + 4q^3(1 - q) + 4q^2(1 - q)^2 = q^2(2 - q^2).$$

Therefore, when K is random and $q = F(t)$, P_k is given by $P_1 = 0$, $P_2 = 1/3$, $P_3 = 1$ and $P_4 = 1$, from (9),

$$R = \frac{1}{3}\binom{4}{2}(1 - q)^2 q^2 + \binom{4}{3}(1 - q)^3 q + (1 - q)^4$$

$$= (1 - q)^2 (1 + 2q - q^2).$$

In addition, setting that

$$(1 - q)^2 (1 + 2q - q^2) = 1 - q,$$

we have $1 - q = (\sqrt{5} - 1)/2 \approx 0.618$. Thus, if $p > 0.618$, then the system reliability is greater than 1-unit system.

Furthermore, when $q = 1 - \exp(-\lambda t)$, MTTF is

$$\mu_{4,P} = \int_0^\infty e^{-2\lambda t}\left(2 - e^{-2\lambda t}\right) dt = \frac{3}{4\lambda}$$

4.2 Example 2

Consider a simple bridge system with four units in Fig. 2, such as a two-terminal logical network, where when two units have failed, if units 1 and 2, 1 and 4, 2 and 3, or 3 and 4 are operative, the system is operative. When each unit has the same reliability as that of Example 1, the system reliability is

$$R = (1 - q_1)(1 - q_2)(1 - q_3)(1 - q_4) + q_1(1 - q_2)(1 - q_3)(1 - q_4)$$
$$+ (1 - q_1)q_2(1 - q_3)(1 - q_4) + (1 - q_1)(1 - q_2)q_3(1 - q_4)$$
$$+ (1 - q_1)(1 - q_2)(1 - q_3)q_4 + (1 - q_1)(1 - q_2)q_3q_4$$
$$+ q_1(1 - q_2)(1 - q_3)q_4 + (1 - q_1)q_2q_3(1 - q_4)$$
$$+ q_1q_2(1 - q_3)(1 - q_4),$$

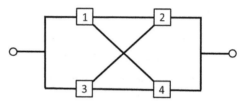

Fig. 2: Bridge system with four units

and the unreliability is

$$1 - R = q_1q_2q_3q_4 + (1 - q_1)q_2q_3q_4 + q_1(1 - q_2)q_3q_4$$
$$+ q_1q_2(1 - q_3)q_4 + q_1q_2q_3(1 - q_4) + q_1(1 - q_2)q_3(1 - q_4)$$
$$+ (1 - q_1)q_2(1 - q_3)q_4.$$

When $q_i = q$,

$$R = (1 - q)^4 + 4q(1 - q)^3 + 4q^2(1 - q)^2 = (1 - q^2)^2,$$
$$1 - R = q^4 + 4q^3(1 - q) + 2q^2(1 - q)^2 = q^2(2 - q^2).$$

Therefore, when K is random and $q = F(t)$, P_k is given by $P_1 = 0$, $P_2 = 2/3$, $P_3 = 1$ and $P_4 = 1$, from (9)

$$R = \frac{2}{3}\binom{4}{2}(1 - q)^2 q^2 + \binom{4}{3}(1 - q)^3 q + (1 - q)^4$$

$$= (1 - q^2)^2.$$

This reliability is greater than that of Example 1. In addition, setting that

$$(1 - q^2)^2 = 1 - q,$$

we have $1 - q = \left(\sqrt{5} + 1\right)/2 \approx 0.382$. Thus, if $p > 0.382$, then the system reliability is greater than that of 1-unit system.

Furthermore, when $q = 1 - \exp(-\lambda t)$, MTTF is

$$\mu_{4,P} = \int_0^\infty \left[1 - \left(1 - e^{-\lambda t}\right)^2 \right]^2 dt = \frac{11}{12\lambda}$$

From the above discussions, we could obtain reliabilities of several kinds of redundant systems, using random K-out-of-N:G systems. If we could derive reliability properties and maintenance policies of random K-out-of-N:G systems theoretically, we could apply these results to actual redundant systems practically by using the above methods.

5. *K*-out-of-*N*:G systems with *i* stages

We propose the following redundant system with i stages: System 1 has an original reliability R_1 and system 2 has its reliability $R_2(R_1)$ which is a function of R_1. Repeating the above procedures, system $(i + 1)$ has its reliability $R_{i+1}(R_i)(i = 1, 2, ...)$, which is called a hierarchical system with i stages.

5.1 2-unit parallel system

As one of the simplest models, we consider a two-unit parallel system with two identical units, which have the same reliability p $(0 < p < 1)$. System 1 has the reliability

$$R_1 = 1 - (1 - p)^2,$$

system 2 has the reliability

$$R_2 = 1 - (1 - p)^4,$$

Table 5: Reliability R_i when $p = 0.65$ where $R_0 \equiv p$

i	0	1	2	3	4	5	6	7	8	9
R_i	0.650	0.666	0.691	0.727	0.778	0.844	0.917	0.975	0.998	1.000

and generally, system i has the reliability

$$R_i = 1 - (1 - p)^{2^i} \ (i = 1, 2, ...),$$

which increases strictly with p from R_1 to 1.

In particular, when $p = \exp(-\lambda t)$, MTTF of the system with i stages is

$$\mu_{2,i} = \int_0^\infty \left[1 - \left(1 - e^{-\lambda t}\right) \right]^{2^i} dt = \frac{1}{\lambda} \sum_{j=1}^{2i} \frac{1}{j}.$$

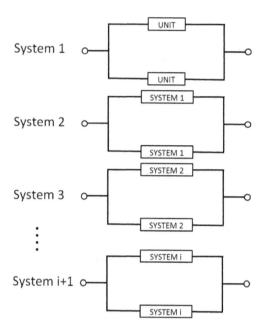

Fig. 3: 2-unit parallel system with *i* stages

5.2 *2-out-of-4 system with two stages*

We consider 2-out-of-4 system with two stages shown in Example 1 of Section 4.1: The configuration of the system is given in Fig. 3. In stage 1, reliability is, from Section 4.1,

$$R_1 = p^2(2 - p^2) .$$

Thus, reliability R_2 in stage 2 is

$$R_2 = p^4 (2 - p^2)^2(2 - p^4 (2 - p^2)^2).$$

When $p = \exp(-\lambda t)$, MTTFs of the systems in stages 1 and 2 are, respectively,

$$\mu_1 = \int_0^\infty e^{-2\lambda t} \left(2 - e^{-2\lambda t}\right) dt = \frac{3}{4\lambda}$$

$$\mu_2 = \int_0^\infty e^{-4\lambda t} \left(2 - e^{-2\lambda t}\right)^2 \left[2 - e^{-4\lambda t} \left(2 - e^{-2\lambda t}\right)^2\right] dt = \frac{1051}{1680\lambda} . \quad (12)$$

Assuming systems with arbitrary stages, their reliability and MTTF can be calculated because their reliability is

$$R_{i+1} = R_i^2 (2 - R_i^2)(i = 1, 2, 3, ...). \quad (1.13)$$

Clearly, R_{i+1} increases strictly with i from $p^2 (2 - p^2)$ to 1 for $p > 0.618$.

Table 5 presents reliability of the system with i stages when $p = 0.65$ where $R_0 \equiv p$. This indicates that R increases strictly with i from 0.65 to 1.

6. Practical applications

When K-out-of-N:G systems are designed, both K and N should be determined in the early designing phase, in which principal specificaitons are confirmed. The fault tolerant computer is an example of such systems and is commonly applied to aircraft flight control systems and nuclear power plant controllers. Number K is determined by considering a minimum constitution which is definitely decided by system design to submit the correct outputs, and N is determined by considering the availability and maintenance cost of the system. When N is large, i.e., the system has enough redundancy, the fault tolerance ability of the system increases; however, the maintenance cost of the redundant subsystem also increases. In the case of nuclear power plant controllers, a large N increases damage tolerance because such systems demand considerably high reliability.

When two sheet metals of an airframe are connected with N rivets, the airframe should be designed to withstand maximum load during operation by K ($K < N$) rivets because some rivets can be broken and lose their strength during flight but the airframe should nonetheless survive in such situations. Airframe structures should also retain their required residual strength for a period of operation after damage has occurred; this principle is called "damagetolerance design". This is required for civil aircraft developments by regulation after the General Dynamics F-111 accident in 1969 and the Dan Air Boeing 707 accident in 1977 (FAR 1978).

Although the damage tolerance design has been inherited until now, additional requirements have been settled by regulations after the Aloha accidents in 1988. After that, Widespread Fatigue Damage (WFD) was firstly introduced to regulations, and full-scale fatigue test certification damage tolerance for WFD was also introducd. The full-scale fatigue test was obligated to certify freedom from WFD for the whole service period (FAA 1998). WFD is defined as the simultaneous presence of cracks at multiple structural details that are of sufficient size and density that the structure will no longer meet its damage tolerance requirement (U.S. Department of Transportation 1998). It was established from 2011 that the Limit of Validity (LOV) and the engineering data which WFD would not occur in airframe structure before it reached LOV, had been required for future civil aircrafts (FAA 2011).

To support LOV in design, parameters should be treated probabilistically because the level of stress is different during each operation and crack growth rates during each operation are also variable.

When we establish a mathematical model of lap joints using K-out-of-N:G system, both K and N may be regarded as stochastic parameters. However, the total number N, which depends on K, should be pre-specified at preflight checks as the number of normal bolts for an inspection standard. For this, one important purpose of K-out-of-N:G system is to determine how many units should be provided according to a stochastic parameter K.

The outer envelope of the airframe is covered with numerous sheet metals that are fastened by rivets. Considering the airframe configuration and its complicated transformation by aerodynamic load changes, the stress on each sheet metal changes greatly and damage tolerances of sheet metals on different portions of the airframe may vary. In such cases, sheet metals are examples of a random K-out-of-N:G system, indicating that there exists a possibility to undergo the efficient maintenance of airframes by considering the reliability of a random K-out-of-N:G system.

It has been supposed, in most parallel systems, that the number N of units is constant and predetermined (Nakagawa 2008). However, because real systems would be complex and large, we might not know the exact number of units of a whole system (Nakagawa and Yun 2011). Such stochastic phenomena arise in order statistics when a sample size is random (Shaked 1975, Rohatgi 1987, Shaked and Wong, 1997, Bartoszewicz 2001, Nanda and Shaked 2008). Generally, the reliability function given in (9) is a general reliability function expressed as

$$h(p) = \sum_{k=1}^{N} \alpha_k h_{N,k}(p), \sum_{k=1}^{N} \alpha_k = 1, \alpha_k \geq 0$$

where $h_{N,k}$ is the reliability function of k-out-of-N:G system and α_k is the convex coefficient (Phillips 1980, Suzuki, Ohi and Kowada 2000, Navarro, Samaniego and Balakrishnan 2011). Thus, we would investigate the reliability properties of random K-out-of-N:G systems, using the reliability $h(p)$, and could apply them to the practical maintenance of some random systems. Reliability quantities, such as MTTF and replacement policies for such parallel systems with a random number of units N were discussed (Nakagawa 2014). Similarly, for K-out-of-N:G system with stochastic parameter K, another purpose is to observe the system reliability $R(F(t))$, in order to estimate mean time to failure (MTTF), and to consider appropriate replacement policies.

REFERENCES

Advisory Circular 25.572C. 1998. Damage tolerance and fatigue evaluation of structure. U.S. Department of Transportation. Federal Aviation Administration, April 29, 1998.

Barlow, B.E. and F. Proschan. 1965. Mathematical Theory of Reliability. John Wiley & Sons, New York.

Barlow, B.E. and F. Proschan. 1975. Statistical Theory of Reliability and Life Testing: Probability Models. Holt, Rinehart and Winston & Sons, New York.

Bartoszewicz, J. 2001. Stochastic comparisons of random minima and maxima from life distributions. Statis. & Prob. Letters 55: 107-112.

Blokus, A. 2006. Reliability analysis of large systems with dependent components. Inter. J. Reliab. Qual. Saf. Eng. 13: 1-14.

Chang, C.J., L. Cui and F.K. Hwang. 2000. Reliability of Consecutive-k Systems. Kluwer, Dordrecht.

FAA. 1998. 14 CFR Part 25 Amendment 25-96.

FAA. 2011. 14 CFR Part 25 Amendment 25-132.

FAR 25.571 Amendment 45, 1978.

Havil, J. 2003. GAMMA: Exploring Euler's Constant. Princeton University Press, Princeton.

Ito, K., X. Zhao and T. Nakagawa. 2017. Random number of unit for k-out-of-n. Applied Mathematics and Computation 45: 563-572.

Kenyon, R.L. and R.L. Newell. 1983. Steady-state availability of k-out-of-n:G system with single repair. IEEE Trans. Rel. R-32: 188-190.

Kuo, W. and M.J. Zuo. 2003. Optimal Reliability Modeling, Principles and Applications. John Wiley & Sons, Hoboken.

Lala, P.K. 1985. Fault Tolerant and Fault Testable Hardware Design. PrenticeHall, London.

Linton, D.G. and J.G. Saw. 1974. Reliability analysis of the k-out-of-n:F system. IEEE Trans. Rel. R-23: 97-103.

MacPherson, A.J. and K.D. Glazebrook. 2011. A dynamic programming policy improvement approach to the development of maintenance policies for 2-phase systems with aging. IEEE Trans. Rel. 60: 448-459.

Nakagawa, T. 1985. Optimization problems in k-out-of-n systems. IEEE Trans. Rel. R-34: 248-250.

Nakagawa, T. 2005. Maintenance Theory of Reliability. Springer, London.

Nakagawa, T. 2008. Advanced Reliability Models and Maintenance Policies. Springer, London.

Nakagawa, T. 2014. Random Maintenance Policies. Springer, London.

Nakagawa, T. and W.Y. Yun. 2011. Note on MTTF of a parallel system. Inter. J. Reliab. Qua. Saf. Eng. 18: 1-8.

Nakagawa, T. and X. Zhao. 2005. Optimization problems of a parallel system with a random number of units. IEEE Trans. on Rel. 61(2): 534-548.

Nanda, K. and M. Shaked. 2008. Partial ordering and aging properties of order statistics when the sample size is random: a brief review. Commun. Statist. Theory Meth. 37: 1710-1720.

Navarro, J., F. Samaniego and B. Balakrishnan. 2011. Signature-based representations for the reliability of systems with heterogeneous components. Journal of Applied Probability 48: 856-867.

Pham, H. 2003. Reliability of system with multiple failure modes. pp. 19-36. *In*: H. Pham (ed.). Handbook of Reliability Engineering. Springer, London.

Phillips, M.J. 1980. *k*-out-of-n:*G* systems are preferable. IEEE Transactions of Reliability R-29(2): 166-169.

Rohatgi, V.K. 1987. Distribution of order statistics with random sample size. Commun. Statis. – Theory Meth. 16: 3739-3743.

Shaked, M. 1975. On the distribution of the minimum and of the maximum of a random number of i.i.d. random variables. pp. 363-380. *In*: G.P. Patil, S. Kots and J. Ord (eds.). Statistical Distributions in Science Work. Reidel Pub., Dordecht, Holland.

Shaked, M. and T. Wong. 1997. Stochastic comparisons of random minima and maxima. J. Appl. Prob. 34: 420-425.

Suzuki, T., F. Ohi and M. Kowada. 2000. Entropy and safety monitoring systems. Japan Journal of Industrial and Applied Mathematics, 17: 59-71.

Ushakov, I.A. 1994. Handbook of Reliability Engineering. John Wiley & Sons, New York.

Wu, S. and M.J. Zuo. 2010. Linear and nonlinear preventive maintenance models. IEEE Trans. Rel. 59: 242-249.

Zia, L. and D.W. Coit. 2010. Redundancy allocation for series-parallel systems using a column generation approach. IEEE Trans. Rel. 59: 706-717.

A Summary of Maintenance Policies for *K*-out-of-*N* Models and Their Applications to Consecutive Systems

Lei Zhou[1*], Taishin Nakamura[1], Toshio Nakagawa[2], Xiao Xiao[1] and Hisashi Yamamoto[1]

[1] Faculty of System Design, Tokyo Metropolitan University, 6-6, Hino, Tokyo, 191-0065, Japan
[2] Department of Business Administration, Aichi Institute of Technology, 1247 Yagusa-cho, Toyota, Aichi, 470-0392, Japan

1. Introduction

K-out-of-*N* models consist of an ordered sequence of *N* units, such that the system is good if and only if at least *K* units in the system are good. Also, a consecutive-*K*-out-of-*N*:G system (Con/*K*/*N*:G system) consists of an ordered sequence of *N* components, such that the system is good if and only if at least *K* consecutive components in the system are good (Kuo and Zuo 2003). With the development of system complexity, high performance, and other requirements, the system reliability also puts forward higher requirements. Reliability has become an important indicator for evaluating the system. In order to reduce the probability of system failure, as well as reduce the damage caused when the failure occurs, proper maintenance and repair can be employed. Performing timely maintenance, repairs and planned preventive maintenance before the failure of a system has become an important topic in the field of reliability research.

This chapter considers the maintenance policies for *K*-out-of-*N* models and their applications to consecutive systems. In section 2, we express the mean time to failure (MTTF) of the standard *K*-out-of-*N* systems and give

*Corresponding author: zhou-lei1@ed.tmu.ac.jp

the maintenance policies: Optimal number of units N^* for given K and replacement time T^* for given K and N are found analytically. In section 3, when K is a random variable, MTTF and reliabilities are obtained, and optimal number N^* and time T^* are also derived. In section 4, we consider the application of maintenance policies to Con/K/N:G system when K is 2. Finally, we find the MTTF, optimal number N^* and time T^* for a consecutive-2-out-of-N:G system.

2. Standard *K*-out-of-*N* systems

Consider a K-out-of-N system ($K = 1, 2, ..., N$ and $N = 1, 2, ...$), for constant K and N, it can be operating if and only if at least K units of N units are operating. It is assumed that each unit has an identical failure distribution $F(t)$ with finite mean $\mu \equiv \int_0^\infty \overline{F}(t) dt < \infty$, a density function $f(t) \equiv dF(t)/dt$, and failure rate $h(t) \equiv f(t)/\overline{F}(t)$, where $\overline{\Phi}(t) \equiv 1 - \Phi(t)$ for any function $\Phi(t)$. Hence, the reliability of the system at time t is (Barlow and Proschan 1965, Nakagawa 2008)

$$R(t) = \sum_{j=0}^{N-K} \binom{N}{j} \left[F(t) \right]^j \left[\overline{F}(t) \right]^{N-j} = \sum_{j=K}^{N} \binom{N}{j} \left[\overline{F}(t) \right]^j \left[F(t) \right]^{N-j}, \quad (1)$$

the mean time to failure (MTTF) is

$$\mu_{N,K} \equiv \int_0^\infty R(t) dt = \sum_{j=0}^{N-K} \binom{N}{j} \int_0^\infty \left[F(t) \right]^j \left[\overline{F}(t) \right]^{N-j} dt,$$

$$= \sum_{j=K}^{N} \binom{N}{j} \int_0^\infty \left[\overline{F}(t) \right]^j \left[F(t) \right]^{N-j} dt, \quad (2)$$

and the failure rate is

$$Q_N(t, K) = -\frac{R'(t)}{R(t)} = \frac{Nh(t) \binom{N-1}{K-1} \left[\overline{F}(t) \right]^K \left[F(t) \right]^{N-K}}{\sum_{j=K}^{N} \binom{N}{j} \left[\overline{F}(t) \right]^j \left[F(t) \right]^{N-j}},$$

$$= \frac{N \binom{N-1}{K-1} h(t)}{\sum_{j=K}^{N} \binom{N}{j} \left[\overline{F}(t)/F(t) \right]^{j-K}}. \quad (3)$$

Thus, because of $\overline{F}(t)/F(t)$ decreases with t to 0, if $h(t)$ increases to $h(\infty) \equiv \lim_{t \to \infty} h(t)$, failure rate $Q_N(t, K)$ increases with t to $Kh(\infty)$.

When $K = N$, the system is called a series system. The reliability, MTTF, and failure rate are

$$R(t) = \left[\overline{F}(t)\right]^N, \tag{4}$$

$$\mu_{N,N} = \int_0^\infty \left[\overline{F}(t)\right]^N dt, \tag{5}$$

$$Q_N(t, N) = Nh(t). \tag{6}$$

When $K = 1$, the system is called a parallel system. The reliability, MTTF and failure rate are

$$R(t) = 1 - [F(t)]^N, \tag{7}$$

$$\mu_{N,1} = \int_0^\infty \left\{1 - \left[F(t)\right]^N\right\} dt, \tag{8}$$

$$Q_N(t, 1) = \frac{Nh(t)\overline{F}(t)\left[F(t)\right]^{N-1}}{1 - \left[F(t)\right]^N}, \tag{9}$$

and when $K = n + 1$ and $N = 2n + 1$, the system is called a majority decision system (Nakagawa 2008). The reliability, MTTF and failure rate are

$$R(t) = \sum_{j=n+1}^{2n+1} \binom{2n+1}{j} \left[\overline{F}(t)\right]^j \left[F(t)\right]^{2n+1-j}, \tag{10}$$

$$\mu_{2n+1,n+1} = \sum_{j=n+1}^{2n+1} \binom{2n+1}{j} \int_0^\infty \left[\overline{F}(t)\right]^j \left[F(t)\right]^{2n+1-j} dt, \tag{11}$$

$$Q_{2n+1}(t, n+1) = \frac{(2n+1)h(t)\binom{2n}{n}\left[\overline{F}(t)\right]^{n+1}\left[F(t)\right]^n}{\sum_{j=n+1}^{2n+1} \binom{2n+1}{j}\left[\overline{F}(t)\right]^j\left[F(t)\right]^{2n+1-j}}. \tag{12}$$

In particular, from (3), when $F(t) = 1 - \exp(-\lambda t)$ (Nakagawa 2014), the failure rate is

$$Q_N(t, K) = \frac{K\lambda\binom{N}{K}e^{-K\lambda t}\left(1 - e^{-\lambda t}\right)^{N-K}}{\sum_{j=K}^{N}\binom{N}{j}e^{-j\lambda t}\left(1 - e^{-\lambda t}\right)^{N-j}}, \tag{13}$$

which increases with t to $K\lambda$ and increases with K to $N\lambda$. The MTTF is

$$\mu_{N,K} = \sum_{j=K}^{N} \binom{N}{j} \int_0^\infty e^{-j\lambda t} \left(1 - e^{-\lambda t}\right)^{N-j} dt = \frac{1}{\lambda} \sum_{j=K}^{N} \frac{1}{j}, \tag{14}$$

and it is approximated by

$$\tilde{\mu}_{N,K} = \frac{1}{\lambda}\left[\ln N - \ln(K-1)\right] = \frac{1}{\lambda} \ln \frac{N}{K-1}, \tag{15}$$

for $K \geq 2$ and large N.

Furthermore, when the failure time is a Weibull distribution $F(t) = 1 - \exp\left[-(\lambda t)^m\right]$ $(m \geq 1)$, MTTF are approximately

$$\tilde{\mu}_{N,K} = \frac{1}{\lambda}\left(\sum_{j=K}^{N} \frac{1}{j}\right)^{1/m}, \tag{16}$$

$$\hat{\mu}_{N,K} = \frac{1}{\lambda}\left(\ln \frac{N}{K-1}\right)^{1/m}. \tag{17}$$

Table 1 presents $\mu_{N,K}$ in (2), $\tilde{\mu}_{N,K}$ in (16) and $\hat{\mu}_{N,K}$ in (17) when $F(t) = 1 - \exp[-(\lambda t)^m]$ $(m = 1, 2, 3)$.

2.1 Number of units

In this subsection, we consider a simple maintenance policy, in which the components are replaced with new ones immediately when the system fails. Under the above maintenance policy, we consider the following costs: Let C_1 be an acquisition cost for one unit and C_R be a replacement cost for a failed system. The expected cost rate until replacement at failure is (Nakagawa 2014)

$$C(N, K) = \frac{NC_1 + C_R}{\mu_{N,K}} \quad (N = K, K+1, \ldots). \tag{18}$$

We find the optimal N^* to minimize $C(N, K)$ for a fixed K. Then, forming the inequality

$$C(N+1, K) - C(N, K) \geq 0, \tag{19}$$

$$\frac{\mu_{N,K}}{\mu_{N+1,K} - \mu_{N,K}} - N \geq \frac{C_R}{C_1}. \tag{20}$$

In particular, when $F(t) = 1 - \exp(-\lambda t)$ is given, from (14) and (20) is

$$(N+1) \sum_{j=K}^{N} \frac{1}{j} - N \geq \frac{C_R}{C_1}, \tag{21}$$

Table 1: MTTF $\mu_{N,K}$, $\tilde{\mu}_{N,K}$ and $\hat{\mu}_{N,K}$ when $F(t) = 1 - \exp[-(\lambda t)^m]$ ($m = 1, 2, 3, \lambda = 1$)

K	m = 1			m = 2			m = 3		
	$\mu_{N,K}$	$\tilde{\mu}_{N,K}$	$\hat{\mu}_{N,K}$	$\mu_{N,K}$	$\tilde{\mu}_{N,K}$	$\hat{\mu}_{N,K}$	$\mu_{N,K}$	$\tilde{\mu}_{N,K}$	$\hat{\mu}_{N,K}$

N = 100

K	$\mu_{N,K}$	$\tilde{\mu}_{N,K}$	$\hat{\mu}_{N,K}$	$\mu_{N,K}$	$\tilde{\mu}_{N,K}$	$\hat{\mu}_{N,K}$	$\mu_{N,K}$	$\tilde{\mu}_{N,K}$	$\hat{\mu}_{N,K}$
1	5.187	5.187	5.187	2.262	2.278	2.277	1.720	1.731	1.731
2	4.187	4.187	4.605	2.037	2.046	2.146	1.606	1.612	1.664
5	3.104	3.104	3.219	1.757	1.762	1.794	1.455	1.459	1.477
10	2.358	2.358	2.408	1.533	1.536	1.552	1.329	1.331	1.340
20	1.640	1.640	1.661	1.278	1.281	1.289	1.177	1.790	1.184
50	0.708	0.708	0.713	0.839	0.842	0.845	0.889	0.891	0.894
60	0.524	0.524	0.528	0.722	0.724	0.726	0.804	0.806	0.808
70	0.369	0.369	0.371	0.605	0.607	0.609	0.715	0.717	0.719
80	0.234	0.234	0.236	0.481	0.484	0.486	0.613	0.617	0.618
90	0.116	0.116	0.117	0.337	0.340	0.341	0.483	0.488	0.488
100	0.010	0.010	0.010	0.089	0.100	0.100	0.192	0.215	0.216

N = 50

K	m = 1			m = 2			m = 3		
	$\mu_{N,K}$	$\tilde{\mu}_{N,K}$	$\hat{\mu}_{N,K}$	$\mu_{N,K}$	$\tilde{\mu}_{N,K}$	$\hat{\mu}_{N,K}$	$\mu_{N,K}$	$\tilde{\mu}_{N,K}$	$\hat{\mu}_{N,K}$
1	4.499	4.499	4.499	2.101	2.121	2.121	1.637	1.651	1.651
2	3.499	3.499	3.912	1.859	1.871	1.978	1.510	1.518	1.576
5	2.416	2.416	2.530	1.548	1.554	1.589	1.337	1.342	1.362
10	1.670	1.670	1.715	1.287	1.292	1.310	1.182	1.186	1.197
20	0.951	0.952	0.968	0.971	0.975	0.984	0.980	0.984	0.989
30	0.538	0.538	0.545	0.729	0.733	0.738	0.809	0.813	0.817
40	0.246	0.246	0.248	0.490	0.496	0.498	0.620	0.626	0.629
50	0.020	0.020	0.020	0.125	0.141	0.142	0.242	0.271	0.272

whose left-side increases with N to ∞. Thus, there exists a finite and unique minimum N^* ($K \leq N^* < \infty$) which satisfies (21) and increases with K.

Also, when $F(t) = 1 - \exp[-(\lambda t)^m]$ is given, using (16) and (17), approximate $\tilde{N}$ and $\hat{N}$ are given, respectively,

$$\frac{\left[\sum_{j=K}^{N}(1/j)\right]^{1/m}}{\left[\sum_{j=K}^{N+1}(1/j)\right]^{1/m} - \left[\sum_{j=K}^{N}(1/j)\right]^{1/m}} - N \geq \frac{C_R}{C_1}, \tag{22}$$

and

$$\frac{\left\{\ln\left[N/(K-1)\right]\right\}^{\frac{1}{m}}}{\left\{\ln\left[(N+1)/(K-1)\right]\right\}^{\frac{1}{m}} - \left\{\ln\left[N/(K-1)\right]\right\}^{\frac{1}{m}}} - N \geq \frac{C_R}{C_1}. \quad (23)$$

Table 2 presents optimal N^*, $\widetilde{N}$ and $\widehat{N}$ given in (20), (22) and (23), respectively, when $F(t) = 1 - \exp[-(\lambda t)^m]$ $(m = 1, 2, 3)$ are given.

Table 2: Optimal N^*, $\widetilde{N}$ and $\widehat{N}$ when $F(t) = 1 - \exp[-(\lambda t)^m]$ $(m = 1, 2, 3)$

	$C_R/C_1 = 50$								
K	m = 1			m = 2			m = 3		
	N^*	$\widetilde{N}$	$\widehat{N}$	N^*	$\widetilde{N}$	$\widehat{N}$	N^*	$\widetilde{N}$	$\widehat{N}$
1	19	19	19	10	10	10	7	7	7
2	26	26	23	14	14	12	10	10	9
5	40	40	39	22	22	21	16	16	16
7	48	48	47	26	26	26	20	20	20
10	59	59	58	33	33	32	25	25	25
20	91	91	90	51	51	51	40	40	40
30	120	120	120	69	69	69	55	55	55
40	149	149	148	86	86	86	69	69	69
	$C_R/C_1 = 100$								
K	m = 1			m = 2			m = 3		
	N^*	$\widetilde{N}$	$\widehat{N}$	N^*	$\widetilde{N}$	$\widehat{N}$	N^*	$\widetilde{N}$	$\widehat{N}$
1	32	32	32	17	17	17	12	12	12
2	42	42	38	22	22	20	16	16	14
5	61	61	59	33	33	32	24	24	23
7	71	71	69	38	38	38	28	28	28
10	84	84	82	46	46	45	34	34	34
20	120	120	119	67	67	66	51	51	51
30	153	153	152	86	86	86	67	67	67
40	184	184	183	104	104	104	82	82	82

2.2 Replacement time

In this subsection, we consider a simple maintenance policy, in which the system is replaced at a planned time T $(0 < T < \infty)$ or at failure, whichever occurs first. Then, the expected cost rate is (Nakagawa 2008), from (1) and (2),

$$C_N(T, K) = \frac{NC_1 + C_R \sum_{j=0}^{K-1} \binom{N}{j} [\bar{F}(T)]^j [F(T)]^{N-j}}{\sum_{j=K}^{N} \binom{N}{j} \int_0^T [\bar{F}(t)]^j [F(t)]^{N-j} dt}.$$ (24)

We find optimal T^* in order to minimize $C_N(T, K)$ for fixed N and K. Differentiating with respect to T and setting it equal to zero,

$$Q_N(T, K) \sum_{j=K}^{N} \binom{N}{j} \int_0^T [\bar{F}(t)]^j [F(t)]^{N-j} dt$$

$$- \sum_{j=0}^{K-1} \binom{N}{j} [\bar{F}(T)]^j [F(T)]^{N-j} = \frac{NC_1}{C_R}.$$ (25)

If $h(t)$ increases strictly to $h(\infty)$, then the left-hand side of (25) increases strictly with T to

$$Kh(\infty)\mu_{N,K} - 1.$$ (26)

Thus, if $h(\infty) > (NC_1 + C_R)/(KC_R\mu_{N,K})$, then there exists a finite and unique T^* $(0 < T^* < \infty)$ which satisfies (25), and the resulting cost rate is

$$C_N(T^*, K) = C_R Q_N(T^*, K).$$ (27)

In particular, when $F(t) = 1 - e^{-\lambda t}$ is given,

$$Q_N(T, K) = \frac{K\lambda \binom{N}{K}}{\sum_{j=K}^{N} \binom{N}{j} (e^{\lambda T} - 1)^{K-j}},$$ (28)

which increases strictly with T from 0 to $K\lambda$ for $K < N$ and is constant $N\lambda$ for $K = N$. Therefore, if $K < N$ and

$$\frac{K}{N} \sum_{j=K+1}^{N} \frac{1}{j} > \frac{C_1}{C_R},$$ (29)

then there exists a finite and unique T^* $(0 < T^* < \infty)$ which satisfies (25). When $K = N$ is given, $Q_N(T, K) = N\lambda$, and $T^* = \infty$.

Table 3 presents optimal T^* in (25) when $F(t) = 1 - \exp[-(\lambda t)^m]$ ($m = 1$, 2, 3), $\lambda = 1$ and $N = 10$.

Table 3: Optimal T^* when $F(t) = 1 - \exp[-(\lambda t)^m]$ $(m = 1, 2, 3)$

K	$C_R/C_1 = 5$			$C_R/C_1 = 10$		
	$m = 1$	$m = 2$	$m = 3$	$m = 1$	$m = 2$	$m = 3$
2	2.461	1.205	1.075	1.575	1.074	1.006
3	1.805	1.042	0.975	1.183	0.927	0.911
4	1.462	0.918	0.892	0.932	0.811	0.830
5	1.263	0.814	0.818	0.752	0.710	0.756
6	1.195	0.722	0.748	0.620	0.619	0.684
7	1.482	0.638	0.679	0.534	0.533	0.612

3. Random number of units

It is assumed that K is a random variable for a specified N $(N = 1, 2, ...)$ and has a probability function $p_{k,N} \equiv \Pr\{K = k\}$ $(k = 1, 2, ..., N)$ (Nakagawa 2014, Ito et al. 2017). Then, the reliability at time t is, from (1),

$$R(t) = \sum_{k=1}^{N} p_{k,N} \sum_{j=0}^{N-k} \binom{N}{j} [F(t)]^j [\overline{F}(t)]^{N-j},$$

$$= \sum_{j=1}^{N} \binom{N}{j} [\overline{F}(t)]^j [F(t)]^{N-j} \sum_{k=1}^{j} p_{k,N}, \tag{30}$$

and the MTTF is

$$\mu_{N,p} = \sum_{j=1}^{N} \left(\sum_{k=1}^{j} p_{k,N} \right) \binom{N}{j} \int_0^\infty [\overline{F}(t)]^j [F(t)]^{N-j} \, dt. \tag{31}$$

In particular, when $F(t) = 1 - e^{-\lambda t}$,

$$\mu_{N,p} = \frac{1}{\lambda} \sum_{j=1}^{N} \frac{1}{j} \sum_{k=1}^{j} p_{k,N} = \frac{1}{\lambda} \sum_{k=1}^{N} p_{k,N} \sum_{j=k}^{N} \frac{1}{j}. \tag{32}$$

From (15), it is approximated by

$$\tilde{\mu}_{N,p} = \frac{1}{\lambda} \sum_{k=1}^{N} p_{k,N} \ln\left(\frac{N}{k-1}\right), \tag{33}$$

for $K \geq 2$ and large N.

We assign a truncated Poisson distribution for $p_{k,N}$. When

$$p_{k,N} = \frac{\theta^{k-1}/(k-1)!}{\sum_{i=0}^{N-1} (\theta^i/i!)} \quad (k = 1, 2, ..., N), \tag{34}$$

with its mean

$$E[K] = \frac{\theta \sum\limits_{i=0}^{N-2} \left(\theta^i/i!\right)}{\sum\limits_{i=0}^{N-1} \left(\theta^i/i!\right)} + 1 , \tag{35}$$

and the MTTF is, from (32),

$$\mu_{N,p} = \frac{1}{\lambda} \frac{\sum\limits_{j=1}^{N} (1/j) \sum\limits_{k=0}^{j-1} \left(\theta^k/k!\right)}{\sum\limits_{j=0}^{N-1} \left(\theta^j/j!\right)} . \tag{36}$$

For large N, $p_{k,N} \approx [\theta^{k-1}/(k-1)!]e^{-\theta}$ with mean $\theta + 1$,

$$\tilde{\mu}_{N,p} = \frac{1}{\lambda} \sum_{j=1}^{N} \frac{1}{j} \sum_{k=0}^{j-1} \frac{\theta^k}{k!} e^{-\theta} . \tag{37}$$

Table 4 presents $\mu_{N,p}$ in (36) and its approximation $\tilde{\mu}_{N,p}$ in (37) when $F(t) = 1 - e^{-t}$, $N = 50, 100$ and $E[K] = w$. MTTF $\mu_{N,p}$ and $\tilde{\mu}_{N,p}$ can be computed as follows: Compute $\tilde{\theta}$ for given w which satisfies

$$\frac{\tilde{\theta} \sum\limits_{i=0}^{N-2} \left(\tilde{\theta}^i/i!\right)}{\sum\limits_{i=0}^{N-1} \left(\tilde{\theta}^i/i!\right)} + 1 = w . \tag{38}$$

Table 4: MTTF $\mu_{N,p}$ and $\tilde{\mu}_{N,p}$ when $F(t) = 1 - e^{-t}$

w	N = 50			N = 100		
	$\tilde{\theta}$	$\mu_{N,p}$	$\tilde{\mu}_{N,p}$	$\tilde{\theta}$	$\mu_{N,p}$	$\tilde{\mu}_{N,p}$
1	0.000	4.499	4.499	0.000	5.187	5.187
2	1.000	3.703	3.703	1.000	4.391	4.391
5	4.000	2.532	2.532	4.000	3.220	3.220
7	6.000	2.130	2.130	6.000	2.818	2.818
10	9.000	1.725	1.725	9.000	2.413	2.413
20	19.000	0.978	0.978	19.000	1.666	1.666
30	29.006	0.555	0.554	29.000	1.243	1.243
40	39.936	0.256	0.238	39.000	0.947	0.947

Table 4 shows $\mu_{N,p}$ and $\tilde{\mu}_{N,p}$, which are computed by (36) and (37), respectively.

Next, we assign a truncated geometric distribution for $p_{k,N}$: When

$$p_{k,N} = \frac{(1-p)p^{k-1}}{1-p^N} \quad (k = 1, 2,..., N), \tag{39}$$

with its mean

$$E[K] = \frac{1}{1-p} - \frac{Np^N}{1-p^N}. \tag{40}$$

When $F(t) = 1 - e^{-\lambda t}$, the MTTF is, from (32),

$$\mu_{N,p} = \frac{1}{\lambda} \frac{\sum\limits_{j=1}^{N}(1/j)(1-p^j)}{1-p^N}. \tag{41}$$

For large N, $p_{k,N} \approx (1-p)p^{k-1}$ with mean $1/(1-p), \tilde{\mu}_{N,p} = \frac{1}{\lambda}\sum\limits_{j=1}^{N}\frac{1}{j}(1-p^j)$.

3.1 Number of units

When K has a probability function $p_{k,N}$ ($k = 1, 2, ..., N$) under the maintenance policy in Subsection 2.1, the expected cost rate is, from (18) and (31),

$$C(N, p) = \frac{NC_1 + C_R}{\sum\limits_{j=1}^{N}\left(\sum\limits_{k=1}^{j}p_{k,N}\right)\binom{N}{j}\int_0^\infty[\bar{F}(t)]^j[F(t)]^{N-j}\,dt} \quad (N = 1, 2, ...). \tag{42}$$

In particular, when $F(t) = 1 - e^{-\lambda t}$, the expected cost rate is, from (32),

$$\frac{C(N, p)}{\lambda} = \frac{NC_1 + C_R}{\sum\limits_{j=1}^{N}(1/j)\sum\limits_{k=1}^{j}p_{k,N}}. \tag{43}$$

Forming the inequality $C(N + 1, p) - C(N, p) \geq 0$, from (20),

$$\frac{\sum\limits_{j=1}^{N}(1/j)\sum\limits_{k=1}^{j}p_{k,N}}{\sum\limits_{j=1}^{N+1}(1/j)\sum\limits_{k=1}^{j}p_{k,N+1} - \sum\limits_{j=1}^{N}(1/j)\sum\limits_{k=1}^{j}p_{k,N}} - N \geq \frac{C_R}{C_1}. \tag{44}$$

Also, when $p_{k,N} = \left[\theta^{k-1}/(k-1)!\right] \Big/ \sum\limits_{i=0}^{N-1}\left(\theta^i/i!\right)$ $(k = 1, 2, ..., N)$, the expected cost rate in (43) is

$$\frac{C(N,p)}{\lambda} = \frac{(NC_1 + C_R)\sum\limits_{i=0}^{N-1}\left(\theta^i/i!\right)}{\sum\limits_{j=1}^{N}(1/j)\sum\limits_{k=0}^{j-1}\left(\theta^k/k!\right)}, \tag{45}$$

and (44) is

$$\frac{\left[\sum\limits_{j=0}^{N}\left(\theta^j/j!\right)\right]\sum\limits_{j=1}^{N}(1/j)\sum\limits_{k=0}^{j-1}\left(\theta^k/k!\right)}{\left[\sum\limits_{j=0}^{N-1}\left(\theta^j/j!\right)\right]\sum\limits_{j=1}^{N+1}(1/j)\sum\limits_{k=0}^{j-1}\left(\theta^k/k!\right) - \left[\sum\limits_{j=0}^{N}\left(\theta^j/j!\right)\right]\sum\limits_{j=1}^{N}(1/j)\sum\limits_{k=0}^{j-1}\left(\theta^k/k!\right)} - N \ge \frac{C_R}{C_1}, \tag{46}$$

whose left-hand side goes to ∞ as $N \to \infty$, and there exists optimal N_p^* $(1 \le N_p^* < \infty)$ which satisfies (46).

When N is large, i.e., $\sum\limits_{i=0}^{N-1}\left(\dfrac{\theta^i}{i!}\right)e^{-\theta} \approx 1$, the MTTF is given in (37), and the asymptotic expected cost rate is

$$\frac{\tilde{C}(N,p)}{\lambda} = \frac{NC_1 + C_R}{\sum\limits_{j=1}^{N}(1/j)\sum\limits_{k=0}^{j-1}\left(\theta^k/k!\right)e^{-\theta}}, \tag{47}$$

and (46) is

$$(N+1)\sum\limits_{j=1}^{N}\frac{1}{j}\sum\limits_{k=0}^{j-1}\frac{\theta^k}{k!}e^{-\theta} - N \ge \frac{C_R}{C_1}, \tag{48}$$

whose left-hand side increases strictly with N to ∞.

Thus, there exists a finite and unique minimum $\widetilde{N}_p(1 \le \widetilde{N}_p < \infty)$ which satisfies (48). Table 5 presents optimal N_p^* in (46) and $\widetilde{N}_p$ in (48) for θ.

When $p_{k,N} = (1-p)p^{k-1}/(1-p^N)$ $(k = 1, 2,..., N)$, the expected cost rate in (43) is

$$\frac{C(N,p)}{\lambda} = \frac{(NC_1 + C_R)\left(1-p^N\right)}{\sum\limits_{j=1}^{N}(1/j)\left(1-p^j\right)}, \tag{49}$$

and (44) is

Table 5: Optimal N_p^* and $\widetilde{N}_p$ when $F(t) = 1 - e^{-t}$ and $E[K] = w$

$w = \theta + 1$	$C_R/C_1 = 50$		$C_R/C_1 = 100$	
	N_p^*	$\widetilde{N}_p$	N_p^*	$\widetilde{N}_p$
1	19	19	32	32
2	24	24	40	40
5	38	38	58	58
7	46	46	69	69
10	57	57	82	82
20	89	89	119	119
30	119	119	151	151

$$\frac{\left(1-p^{N+1}\right)\sum_{j=1}^{N}(1/j)\left(1-p^j\right)}{\left(1-p^N\right)\sum_{j=1}^{N+1}(1/j)\left(1-p^j\right)-\left(1-p^{N+1}\right)\sum_{j=1}^{N}(1/j)\left(1-p^j\right)} - N \geq \frac{C_R}{C_1}. \quad (50)$$

When N is large, the asymptotic expected cost rate is

$$\frac{\tilde{C}(N,p)}{\lambda} = \frac{NC_1 + C_R}{\sum_{j=1}^{N}(1/j)\left(1-p^j\right)}, \quad (51)$$

and

$$(N+1)\sum_{j=1}^{N}\frac{1}{j}\left(1-p^j\right) - N \geq \frac{C_R}{C_1}. \quad (52)$$

3.2 Replacement time

In this subsection, we consider the maintenance policy, in which the system is replaced at a planned time T $(0 < T < \infty)$ or at failure, whichever occurs first (as in Subsection 2.2). When K has a probability function $p_{k,N}$, the expected cost rate in (24) is (Nakagawa 2014, Ito et al. 2017)

$$C_N(T,p) = \frac{NC_1 + C_R \sum_{k=1}^{N} p_{k,N} \sum_{j=0}^{k-1} \binom{N}{j}\left[\overline{F}(T)\right]^j \left[F(T)\right]^{N-j}}{\sum_{k=1}^{N} p_{k,N} \sum_{j=k}^{N} \binom{N}{j} \int_0^T \left[\overline{F}(t)\right]^j \left[F(t)\right]^{N-j} dt}. \quad (53)$$

Differentiating $C_N(T, p)$ with respect to T and setting it equal to zero,

$$Q_N(T,p)\sum_{k=1}^{N}p_{k,N}\sum_{j=k}^{N}\binom{N}{j}\int_0^T\left[\overline{F}(t)\right]^j\left[F(t)\right]^{N-j}dt$$

$$-\sum_{k=1}^{N}p_{k,N}\sum_{j=0}^{k-1}\binom{N}{j}\left[\overline{F}(T)\right]^j\left[F(T)\right]^{N-j}=\frac{NC_1}{C_R},\tag{54}$$

where

$$Q_N(T,p)\equiv\frac{Nh(T)\sum_{k=1}^{N}p_{k,N}\binom{N-1}{k-1}\left[\overline{F}(T)\right]^k\left[F(T)\right]^{N-k}}{\sum_{k=1}^{N}p_{k,N}\sum_{j=k}^{N}\binom{N}{j}\left[\overline{F}(T)\right]^j\left[F(T)\right]^{N-j}}.\tag{55}$$

When $p_{k,N}=\left[\theta^{k-1}/(k-1)!\right]\Bigg/\sum_{i=0}^{N-1}\left(\theta^i/i!\right)$ $k=1,2,...,N)$ and $F(t)=1-e^{-\lambda t}$, the expected cost rate in (53) is

$$C_N(T,p)=\frac{NC_1\sum_{k=0}^{N-1}\left(\theta^k/k!\right)+C_R\sum_{k=0}^{N-1}\left(\theta^k/k!\right)\sum_{j=0}^{k}H_j(T)}{\sum_{k=0}^{N-1}\left(\theta^k/k!\right)\sum_{j=k+1}^{N}\int_0^T H_j(t)dt},\tag{56}$$

and (54) is

$$\frac{\lambda\sum_{k=0}^{N-1}\left[(k+1)\theta^k/k!\right]H_{k+1}(T)}{\sum_{k=0}^{N-1}\left(\theta^k/k!\right)\sum_{j=k+1}^{N}H_j(T)}\sum_{k=0}^{N-1}\frac{\theta^k}{k!}\sum_{j=k+1}^{N}\int_0^T H_j(t)dt$$

$$-\sum_{k=0}^{N-1}\frac{\theta^k}{k!}\sum_{j=0}^{k}H_j(T)=\frac{NC_1}{C_R}\sum_{k=0}^{N-1}\frac{\theta^k}{k!},\tag{57}$$

where

$$H_j(T)\equiv\binom{N}{j}\left(e^{-\lambda T}\right)^j\left(1-e^{-\lambda T}\right)^{N-j}\quad(j=0,1,...,N).\tag{58}$$

For large N, when $p_{k,N}\approx[\theta^{k-1}/(k-1)!]e^{-\theta}$ and $H_j(T)\approx\left[\left(Ne^{-\lambda T}\right)^j\big/j!\right]$ $\exp(-Ne^{-\lambda T})$, and from (57), approximate $\tilde{T}_p$ satisfies

$$\frac{\lambda \sum_{k=0}^{N-1}\left[(k+1)\theta^k/k!\right]\tilde{H}_{k+1}(T)}{\sum_{k=0}^{N-1}\left(\theta^k/k!\right)\sum_{j=k+1}^{N}\tilde{H}_j(T)}\sum_{k=0}^{N-1}\frac{\theta^k}{k!}e^{-\theta}\sum_{j=k+1}^{N}\int_0^T\tilde{H}_j(t)dt$$

$$-\sum_{k=0}^{N-1}\frac{\theta^k}{k!}e^{-\theta}\sum_{j=0}^{k}\tilde{H}_j(T)=\frac{NC_1}{C_R}, \tag{59}$$

where

$$\tilde{H}_j(T)\equiv\frac{\left(Ne^{-\lambda T}\right)^j}{j!}\exp\left(-Ne^{-\lambda T}\right)\quad(j=0,1,...,N). \tag{60}$$

Table 6 presents optimal T^* in (25), T_p^* in (57) and $\tilde{T}_p$ in (59), when $F(t) = 1 - e^{-t}$, which are obtained as follows:

(i) Optimal T^* satisfies

$$\frac{K\lambda\binom{N}{K}}{\sum_{j=K}^{N}\binom{N}{j}\left(e^{\lambda T^*}-1\right)^{K-j}}\sum_{j=K}^{N}\binom{N}{j}\int_0^{T^*}\left(e^{-\lambda t}\right)^j\left(1-e^{-\lambda t}\right)^{N-j}dt$$

$$-\sum_{j=0}^{K-1}\binom{N}{j}\left(e^{-\lambda T^*}\right)^j\left(1-e^{-\lambda T^*}\right)^{N-j}=\frac{NC_1}{C_R}. \tag{61}$$

(ii) Optimal T_p^* is computed as: First, compute $\tilde{\theta}$ which satisfies, for given w,

$$\frac{\tilde{\theta}\sum_{i=0}^{N-2}\left(\tilde{\theta}^i/i!\right)}{\sum_{i=0}^{N-1}\left(\tilde{\theta}^i/i!\right)}+1=w, \tag{62}$$

and when $w = 1$, $\tilde{\theta} = 0$. Compute T_p^* which satisfies (57) for $\tilde{\theta}$.

(iii) Approximate $\tilde{T}_p$ is computed as: When $\tilde{\theta} = w - 1$, compute $\tilde{T}_p$ which satisfies (59) for $\tilde{\theta}$.

Next, when $p_{k,N} = (1-p)p^{k-1}/(1-p^N)$ ($k = 1, 2,..., N$) and $F(t) = 1 - e^{-\lambda t}$, the expected cost rate in (53) is

Table 6: Optimal T_p^*, and $\widetilde{T}_p$ when $F(t) = 1 - e^{-t}$ and $N = 100$

w	$C_R/C_1 = 50$			$C_R/C_1 = 100$		
	T^*	T_p^*	$\widetilde{T}_p$	T^*	T_p^*	$\widetilde{T}_p$
1	4.48	4.48	4.50	3.87	3.90	3.87
2	3.59	3.78	3.80	3.25	3.23	3.24
5	2.68	2.74	2.76	2.49	2.44	2.44
7	2.37	2.41	2.43	2.22	2.16	2.16
10	2.05	2.08	2.09	1.93	1.87	1.87
20	1.43	1.45	1.47	1.35	1.30	1.31
30	1.07	1.09	1.12	1.01	0.97	0.98
40	0.81	0.85	0.89	0.89	0.74	0.75

$$C_N(T,p) = \frac{NC_1\left(1-p^N\right) + C_R \sum_{k=0}^{N-1}\left(p^k - p^N\right)H_k(T)}{\sum_{k=1}^{N}\left(1-p^k\right)\int_0^T H_k(t)\,dt}, \tag{63}$$

and (54) is

$$\frac{\lambda(1-p)\sum_{k=0}^{N-1}(k+1)p^k H_{k+1}(T)}{\sum_{k=1}^{N}\left(1-p^k\right)H_k(T)} \quad \sum_{k=1}^{N}\left(1-p^k\right)\int_0^T H_k(t)\,dt$$

$$- \sum_{k=0}^{N-1}\left(p^k - p^N\right)H_k(T) = \frac{NC_1}{C_R}\left(1-p^N\right). \tag{64}$$

For large N, when $p_k = (1-p)p^{k-1}$, approximate $\widetilde{T}_p$ satisfies

$$\frac{\lambda(1-p)\sum_{k=0}^{N-1}(k+1)p^k H_{k+1}(T)}{\sum_{k=1}^{N}\left(1-p^k\right)H_k(T)} \quad \sum_{k=1}^{N}\left(1-p^k\right)\int_0^T H_k(t)\,dt$$

$$- \sum_{k=0}^{N-1} p^k H_k(T) = \frac{NC_1}{C_R}. \tag{65}$$

4. Consecutive-2-out-of-*N*:G system

A consecutive-K-out-of-N:F system, abbreviated as Con/K/N:F system, consists of N linearly ordered components, and the system fails if and only if at least K consecutive components fail. In contrast, a consecutive-K-out-of-N:G system, abbreviated as Con/K/N:G system, consists of N linearly ordered components and the system functions if and only if at least K consecutive components function (Kuo and Zuo 2003). Both systems are the duals of each other.

The problem of evaluating the reliability of the Con/K/N system has been discussed by many papers (Hwang 1982, Bollinger and Salvia 1982, Bollinger and Salvia 1985, Papastavridis and Hadjichristos 1987, Kossow and Preuss 1991, Sasaki et al,. 1994, Tong 1985, Gera 2000, Kuo et al. 1990, Zuo and Kuo 1990). Hwang (1982) gave recursive equations for the reliability of a Con/K/N:F system with nonidentical component reliabilities. Bollinger and Salvia (1982, 1985) provided formulas and algorithms for recursive reliability computation. They also gave MTTF of a Con/K/N:F system with components having an exponential life distribution. Papastavridis and Hadjichristos (1987) obtained a general formula for MTTF of a Con/K/N:F system with identical component-failure probabilities, and applied the result to such a system with components having a Weibull life distribution. Kossow and Preuss (1991) presented new simple formulas for MTTF of a Con/K/N:F system and gave special cases with components having Weibull and exponential life distribution. Sasaki et al. (1994) presented the mean time between failures for a repairable Con/K/N:F system by using the frequency of system failure and availability. Tong (1985) reported the first study on the Con/K/N:G system. Gera (2000) used a matrix-formed system reliability and reconsidered the reliability of linear Con/K/N:G system. Furthermore, Kuo et al. (1990) considered the optimal arrangement problem of the Con/K/N:G system. Zuo and Kuo (1990) also proved the invariant optimal assignment for the Con/K/N:G cycle when $2K + 1 \geq N$.

We are interested in MTTF, the optimal number N^* of components and replacement time T^* for Con/K/N:G system. It is assumed that all components have independent and identical life distributions, of which the reliability at time t is $\bar{F}(t)$, with finite mean $\mu \equiv \int_0^\infty \bar{F}(t)\, dt < \infty$, a density function $f(t) \equiv dF(t)/dt$, and failure rate $h(t) \equiv f(t)/\bar{F}(t)$. $[a]$ denotes the largest integer less than or equal to a, $R_G(2, N)$ denotes the reliability of the Con/K/N:G system. Then

$R_G(2, N) = \Pr\{\text{the system is functioning}\},$

$$= \sum_{j=0}^{[(N+1)/2]} \Pr\{\, j \text{ components are functioning, and the system is functioning}\}$$

$$+ \sum_{j=[(N+3)/2]}^{N} \Pr\{\, j \text{ components are functioning, and the system is functioning}\}. \qquad (66)$$

If the number of functioning components is greater than $[(N + 1)/2]$, then there exist two consecutive functioning components in the system, i.e., the system functions. Hence, the right part of the sum

$$\sum_{j=[(N+3)/2]}^{N} \text{Pr}\{j \text{ components are functioning, and the system is functioning}\}$$

$$= \sum_{j=[(N+3)/2]}^{N} \binom{N}{j} \bar{F}(t)^{j} F(t)^{N-j} . \tag{67}$$

If j components are functioning, $j \leq [(N + 1)/2]$, the system fails if there is at least one failed component between every two functioning components. As Fig. 1 shows there is a lattice between each failed component, and there are $(N - j + 1)$ lattices for j functioning components to choose in random.

So, the number of such combinations between failed and functioning components is $\binom{N - j + 1}{j}$; thus, the number of combinations that if j components are functioning then the system functions is $\binom{N}{j} - \binom{N - j + 1}{j}$. Also, when $j = 0$ and $j = 1$, the system indeed fails. The reliability of the system is

$$\sum_{j=2}^{[(N+1)/2]} \left[\binom{N}{j} - \binom{N - j + 1}{j} \right] \bar{F}(t)^{j} F(t)^{N-j} . \tag{68}$$

Overall, the reliability of the system at time t is

$$R_G(2, N) = \sum_{j=2}^{[(N+1)/2]} \left[\binom{N}{j} - \binom{N - j + 1}{j} \right] \bar{F}(t)^{j} F(t)^{N-j}$$

$$+ \sum_{j=[(N+3)/2]}^{N} \binom{N}{j} \bar{F}(t)^{j} F(t)^{N-j} ,$$

$$= \sum_{j=2}^{N} \binom{N}{j} \bar{F}(t)^{j} F(t)^{N-j} - \sum_{j=2}^{[(N+1)/2]} \binom{N - j + 1}{j} \bar{F}(t)^{j} F(t)^{N-j} . \tag{69}$$

By binomial theorem,

$$\sum_{j=0}^{N} \binom{N}{j} \bar{F}(t)^{j} F(t)^{N-j} = \left(\bar{F}(t) + F(t) \right)^{N} = 1 , \tag{70}$$

holds and for $j = 0, 1$,

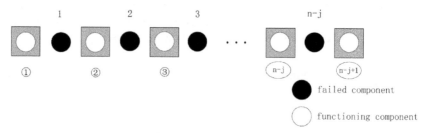

failed component

functioning component

Fig. 1: No-linked functioning components.

$$\binom{N}{j} = \binom{N-j+1}{j}, \tag{71}$$

hence, we can obtain

$$R_F(2, N) = 1 - \sum_{j=0}^{[(N+1)/2]} \binom{N-j+1}{j} \overline{F}(t)^j F(t)^{N-j}. \tag{72}$$

Chiang and Niu (1981) developed an equation for the reliability of a Con/2/N:F system as follows:

$$R_F(2, N) = 1 - \sum_{j=0}^{[(N+1)/2]} \binom{N-j+1}{j} \overline{F}(t)^j F(t)^{N-j}. \tag{73}$$

Furthermore, Kuo et al. (1990) clarified the relationship between Con/K/N:F system and Con/K/N:G system.

Lemma (Kuo et al. 1990)

If the reliability of component i in one type of the Con/K/N:F (G) system is equal to the unreliability of component i in the other type of Con/K/N:G (F) system for $i = 1, 2, \ldots, N$, given that both types of systems have the same values of N and K, then the reliability of one type of system is equal to the unreliability of the other type of the system.

According to the equation of Chiang and Niu (1981) and Lemma of Kuo et al. (1990), the reliability of Con/2/N:G system can be expressed as

$$R_F(2, N) = 1 - \sum_{j=0}^{[(N+1)/2]} \binom{N-j+1}{j} \overline{F}(t)^j F(t)^{N-j}, \tag{74}$$

namely, we can obtain the same result.

The mean time to failure (MTTF) of the system is

$$\mu_{N,2} = \int_0^\infty R_G(2, N)\,dt,$$

$$= \sum_{j=2}^{N}\binom{N}{j}\int_0^\infty \bar{F}(t)^j F(t)^{N-j}\,dt - \sum_{j=2}^{[(N+1)/2]}\binom{N-j+1}{j}\int_0^\infty \bar{F}(t)^j F(t)^{N-j}\,dt,$$

$$= \int_0^\infty \left[1 - \sum_{j=0}^{[(N+1)/2]}\binom{N-j+1}{j}\bar{F}(t)^j F(t)^{N-j}\right]dt, \tag{75}$$

and failure rate is

$$Q_N(t, 2) = -\frac{R'_G(2, N)}{R_G(2, N)},$$

$$= \frac{\left[\begin{array}{l}\displaystyle\sum_{j=1}^{[(N-1)/2]}(2N-3j+1)\binom{N-j}{j-1}\bar{F}(t)^j F(t)^{N-j-1} \\[2mm] \displaystyle +\binom{N-[\frac{N-1}{2}]}{[\frac{N+1}{2}]}\left(N-[\tfrac{N+1}{2}]\right)\bar{F}(t)^{[\frac{N+1}{2}]}F(t)^{N-[\frac{N+3}{2}]}\end{array}\right]h(t)}{1/\bar{F}(t) - \displaystyle\sum_{j=0}^{[\frac{N+1}{2}]}\binom{N-j+1}{j}\bar{F}(t)^{j-1}F(t)^{N-j}},$$

$$= \begin{cases} \dfrac{\left[\begin{array}{l}\displaystyle\sum_{j=1}^{(N-1)/2}(2N-3j+1)\binom{N-j}{j-1}\bar{F}(t)^j F(t)^{N-j-1} \\[2mm] \displaystyle +\frac{N-1}{2}\bar{F}(t)^{\frac{N+1}{2}}F(t)^{\frac{N-3}{2}}\end{array}\right]h(t)}{1/\bar{F}(t) - \displaystyle\sum_{j=0}^{(N+1)/2}\binom{N-j+1}{j}\bar{F}(t)^{j-1}F(t)^{N-j}}, & \text{if } N = 2k+1, \\[12mm] \dfrac{\left[\begin{array}{l}\displaystyle\sum_{j=1}^{(N-2)/2}(2N-3j+1)\binom{N-j}{j-1}\bar{F}(t)^j F(t)^{N-j-1} \\[2mm] \displaystyle +\frac{N^2+2N}{4}\bar{F}(t)^{\frac{N}{2}}F(t)^{\frac{N-2}{2}}\end{array}\right]h(t)}{1/\bar{F}(t) - \displaystyle\sum_{j=0}^{N/2}\binom{N-j+1}{j}\bar{F}(t)^{j-1}F(t)^{N-j}}, & \text{if } N = 2k, \end{cases} \tag{76}$$

$(k = 1, 2, 3...)$

In particular, when $F(t) = 1 - e^{-\lambda t}$ (exponential lifetimes), MTTF is

$$\mu_{N,2} = \sum_{j=2}^{N} \binom{N}{j} \int_0^\infty e^{-j\lambda t}(1 - e^{-\lambda t})^{N-j} dt$$

$$- \sum_{j=2}^{[(N+1)/2]} \binom{N-j+1}{j} \int_0^\infty e^{-j\lambda t}(1 - e^{-\lambda t})^{N-j} dt,$$

$$= \frac{1}{\lambda}\sum_{j=2}^{N}\frac{1}{j} - \frac{1}{\lambda}\sum_{j=2}^{[(N+1)/2]}\frac{\binom{N-j+1}{j}}{j \cdot \binom{N}{j}}, \tag{77}$$

$$= \frac{1}{\lambda}\sum_{j=2}^{N}\frac{1}{j} - \frac{1}{\lambda}\sum_{j=2}^{[(N+1)/2]}\frac{1}{j}\prod_{l=0}^{j-1}\left(1 - \frac{j-1}{N-l}\right)$$

from (75).

Furthermore, when the failure time has Weibull distribution $F(t) = 1 - \exp[-(\lambda t)^m]$ ($m \geq 1$), MTTF is obtained approximately

$$\hat{\mu}_{N,2} = \sum_{j=2}^{N} \binom{N}{j} \int_0^\infty e^{-j(\lambda t)^m}(1 - e^{-(\lambda t)^m})^{N-j} dt$$

$$- \sum_{j=2}^{[(N+1)/2]} \binom{N-j+1}{j} \int_0^\infty e^{-j(\lambda t)^m}(1 - e^{-(\lambda t)^m})^{N-j} dt, \tag{78}$$

$$\approx \frac{1}{\lambda}\left(\sum_{j=2}^{N}\frac{1}{j}\right)^{1/m} - \frac{1}{\lambda}\left[\sum_{j=2}^{[(N+1)/2]}\frac{\binom{N-j+1}{j}}{j \cdot \binom{N}{j}}\right]^{1/m},$$

$$= \frac{1}{\lambda}\left(\sum_{j=2}^{N}\frac{1}{j}\right)^{1/m} - \frac{1}{\lambda}\left(\sum_{j=2}^{[(N+1)/2]}\frac{1}{j}\prod_{l=0}^{j-1}\left(1 - \frac{j-1}{N-l}\right)\right)^{1/m}.$$

Table 7 presents $\mu_{N,2}$ in (75) and $\hat{\mu}_{N,2}$ in (78) when $F(t) = 1 - \exp[-(\lambda t)^m]$ ($m = 1, 2, 3$).

4.1 Number of units

Using the same maintenance policy as in Subsection 2.1, we let C_1 be an acquisition cost of one unit and C_R be a replacement cost of a failed system. The expected cost rate until replacement at failure is

Table 7: MTTF $\mu_{N,2}$ and $\hat{\mu}_{N,2}$ when $F(t) = 1 - \exp[-(\lambda t)^m]$ $(m = 1, 2, 3)$

N	m = 1		m = 2		m = 3	
	$\mu_{N,2}$	$\hat{\mu}_{N,2}$	$\mu_{N,2}$	$\hat{\mu}_{N,2}$	$\mu_{N,2}$	$\hat{\mu}_{N,2}$
3	0.667	0.667	0.742	0.505	0.798	0.391
4	0.833	0.833	0.857	0.541	0.888	0.397
5	0.950	0.950	0.925	0.556	0.937	0.393
10	1.327	1.327	1.119	0.613	1.071	0.400
20	1.701	1.701	1.281	0.665	1.175	0.410
50	2.187	2.187	1.463	0.725	1.286	0.423
60	2.283	2.283	1.496	0.736	1.305	0.426
70	2.364	2.364	1.523	0.746	1.321	0.428
80	2.433	2.433	1.547	0.754	1.335	0.430
90	2.495	2.495	1.567	0.760	1.347	0.432
100	2.549	2.549	1.584	0.767	1.357	0.433

$$C(N, 2) = \frac{NC_1 + C_R}{\mu_{N,2}} \quad (N = 2, 3, ...) . \tag{79}$$

We find the optimal N^* to minimize $C(N, 2)$. Then, forming the inequality $C(N + 1, 2) - C(N, 2) \geq 0$, we have:

$$\frac{(N + 1)C_1 + C_R}{\mu_{N+1,2}} - \frac{NC_1 + C_R}{\mu_{N,2}} \geq 0, \tag{80}$$

that is

$$\frac{\mu_{N,2}}{\mu_{N+1,2} - \mu_{N,2}} - N \geq \frac{C_R}{C_1} . \tag{81}$$

In particular, when $F(t) = 1 - e^{-\lambda t}$, from (77), and (81) is

$$\frac{\sum_{j=2}^{N} \frac{1}{j} - \sum_{j=2}^{[(N+1)/2]} \frac{1}{j} \prod_{l=0}^{j-1} \left(1 - \frac{j-1}{N-l}\right)}{\frac{1}{N+1} - \left(\sum_{j=2}^{[(N+2)/2]} \frac{1}{j} \prod_{l=0}^{j-1} \left(1 - \frac{j-1}{N+1-l}\right) - \sum_{j=2}^{[(N+1)/2]} \frac{1}{j} \prod_{l=0}^{j-1} \left(1 - \frac{j-1}{N-l}\right)\right)} - N \geq \frac{C_R}{C_1} . \tag{82}$$

$$
\left\{
\begin{array}{l}
\dfrac{\displaystyle\sum_{j=2}^{N}\dfrac{1}{j}-\sum_{j=2}^{(N+1)/2}\dfrac{1}{j}\prod_{l=0}^{j-1}\left(1-\dfrac{j-1}{N-l}\right)}{1-\displaystyle\sum_{j=2}^{(N+1)/2}\dfrac{j-1}{N-2j+2}\prod_{l=0}^{j-1}\left(1-\dfrac{j-1}{N-l}\right)}(N+1)-N\ge\dfrac{C_R}{C_1}, \\[4ex]
\hfill N=2k+1, \\[3ex]
\dfrac{\left[\displaystyle\sum_{j=2}^{N}\dfrac{1}{j}-\sum_{j=2}^{N/2}\dfrac{1}{j}\prod_{l=0}^{j-1}\left(1-\dfrac{j-1}{N-l}\right)\right](N+1)(N+2)}{(N+2)-\left[\begin{array}{l}(N+2)\displaystyle\sum_{j=2}^{N/2}\dfrac{j-1}{N-2j+2}\prod_{l=0}^{j-1}\left(1-\dfrac{j-1}{N-l}\right)\\[2ex]+2(N+1)\displaystyle\prod_{l=0}^{N/2}\left(1-\dfrac{N}{2(N+1-l)}\right)\end{array}\right]}-N\ge\dfrac{C_R}{C_1}, \quad (83)\\[6ex]
\hfill N=2k,
\end{array}
\right.
$$

$(k=1,2,3...)$

whose left-side increases with N to ∞. Thus, there exists a finite and unique minimum N^* $(2\le N^*<\infty)$ which satisfies (83).

In addition, when $F(t)=1-\exp[-(\lambda t)^m]$, using (78), approximate $\widetilde{N}$ is given,

$$
\dfrac{\left(\displaystyle\sum_{j=2}^{N}\dfrac{1}{j}\right)^{1/m}-\left(\displaystyle\sum_{j=2}^{[(N+1)/2]}\dfrac{1}{j}\prod_{l=0}^{j-1}\left(1-\dfrac{j-1}{N-l}\right)\right)^{1/m}}{\dfrac{1}{(N+1)^{1/m}}-\left(\displaystyle\sum_{j=2}^{[(N+2)/2]}\dfrac{1}{j}\prod_{l=0}^{j-1}\left(1-\dfrac{j-1}{N+1-l}\right)\right)^{1/m}}-N\ge\dfrac{C_R}{C_1}. \quad (84)
$$

$$
+\left(\displaystyle\sum_{j=2}^{[(N+1)/2]}\dfrac{1}{j}\prod_{l=0}^{j-1}\left(1-\dfrac{j-1}{N-l}\right)\right)^{1/m}
$$

Table 8 presents optimal N^* and $\widetilde{N}$ given in (81) and (84), respectively, when $F(t)=1-\exp[(-\lambda t)^m]$ $(m=1,2,3)$.

4.2 Replacement time

Suppose that the system is replaced at a planned time T $(0<T<\infty)$ or at failure, whichever occurs first; the expected cost rate is then

$$
C_N(T,2)=\dfrac{NC_1+C_R(1-R_G(T))}{\displaystyle\int_0^T R_G(t)dt},
$$

Table 8: Optimal N^* and $\tilde{N}$ when $F(t) = 1 - \exp[-(\lambda t)^m]$ $(m = 1, 2, 3)$

C_R/C_1	$m = 1$		$m = 2$		$m = 3$	
	N^*	$\tilde{N}$	N^*	$\tilde{N}$	N^*	$\tilde{N}$
30	16	16	4	4	3	3
50	22	22	8	8	3	3
80	30	30	11	11	5	5
100	35	35	13	13	5	5
200	60	60	25	25	5	5
300	80	80	30	30	6	6
400	100	100	40	40	15	15
500	120	120	48	48	18	18

$$
= \frac{NC_1 + C_R \displaystyle\sum_{j=0}^{[(N+1)/2]} \binom{N-j+1}{j} \bar{F}(T)^j F(T)^{N-j}}{T - \displaystyle\sum_{j=0}^{[(N+1)/2]} \binom{N-j+1}{j} \int_0^T \bar{F}(t)^j F(t)^{N-j}\,dt}, \tag{85}
$$

from (72). We find the optimal T^* to minimize $C_N(T, 2)$. Differentiating $C_N(T, 2)$ with respect to T and setting it equal to zero,

$$
C_N'(T, 2) = \frac{-C_R R_G'(T) \int_0^T R_G(t)dt - R_G(T)(NC_1 + C_R(1 - R_G(T)))}{\left(\int_0^T R_G(t)dt\right)^2} = 0, \tag{86}
$$

namely,

$$
-\frac{R_G'(T)}{R_G(T)} C_R \int_0^T R_G(t)dt = NC_1 + C_R(1 - R_G(T)), \tag{87}
$$

and we let

$$
1 - R_G(T) = F_G(T), \tag{88}
$$

so, we have

$$
Q_N(T, 2) \cdot \int_0^T R_G(t)dt - F_G(T) = \frac{NC_1}{C_R}. \tag{89}
$$

When $F(t) = 1 - e^{-\lambda t}$ (exponential lifetimes), our numerical experiment shows that if $h(t)$ increases strictly to $h(\infty)$, then the left-hand side of (89) increases strictly with T to $2^*\lambda\mu_{N,2} - 1$. Thus, we speculate that if $2^*\lambda\mu_{N,2} - 1 > \frac{NC_1}{C_R}$, there exists a finite and unique T^* $(0 < T^* < \infty)$ which satisfies (85). The resulting cost rate is

$$
C_N(T^*, 2) = C_R \cdot Q_N(T^*, 2). \tag{90}
$$

Table 9: Optimal T^* when $F(t) = 1 - e^{-\lambda t}$

N	$C_R/C_1 = 10$	$C_R/C_1 = 50$
5	0.944	0.292
6	0.997	0.349
7	1.078	0.413
8	1.164	0.471
9	1.251	0.527
10	1.338	0.580
12	1.510	0.677
15	1.771	0.805

Table 9 presents optimal T^* in (89) when $F(t) = 1 - e^{-\lambda t}$, $\lambda = 1$.

5. Conclusion

In this chapter, we considered the optimization policies in K-out-of-N models, in two cases when K are defined constantly and randomly, and their applications to Con/2/N:G systems. In K-out-of-N models, the system's MTTF and reliabilities were obtained, respectively. The optimal number N^* of units and replacement time T^* were also analytically given, respectively. Furthermore, we derived optimal results for Con/2/N:G systems by using the same maintenance policies.

Obviously, it is a practical problem to consider a Con/2/N:G system, such as street parking systems. In the future, we will consider for Con/K/N:G system for $N \geq K$ and obtain MTTF and reliabilities and also try to derive optimal policies for such complex systems, using the same maintenance results.

REFERENCES

Barlow, R.E. and F. Proschan. 1965. Mathematical Theory of Reliability. Wiley, New York.

Bollinger, R.C. and A.A. Salvia. 1982. Consecutive-k-out-of-n:F networks. IEEE Trans. Reliability R-31(1): 53-56.

Bollinger, R.C. and A.A. Salvia. 1985. Consecutive-k-out-of-n:F systems with sequential failures. IEEE Trans. Reliability R-34: 43-45.

Chiang, D.T. and S.C. Niu. 1981. Reliability of a consecutive-k-out-of-n:F system. IEEE Trans. Reliability 30(1): 87-89.

Gera, A.E. 2000. A consecutive k-out-of-n:G system with dependent elements – a matrix formulation and solution. Reliab. Engng Syst. Saf. 68: 61-67.

Hwang, F.K. 1982. Fast solutions for consecutive-k-out-of-n:F systems. IEEE Trans. Reliability R-31: 447-448.

Ito, K. et al. 2017. Random number of units for K-out-of-n systems. Applied Mathematical Modelling 45: 563-572.

Kossow, A. and W. Preuss. 1991. Mean time-to-failure for a linear-consecutive-k-out-of-n:F system. IEEE Trans. Reliability 40(3): 271-272.

Kuo, W. et al. 1990. A consecutive-k-out-of-n:G system: the mirror image of a consecutive-k-out-of-n:F system. IEEE Trans. Reliability 39(2): 244-253.

Kuo, W. and M.J. Zuo. 2003. Optimal Reliability Modeling. Wiley, New York.

Nakagawa, T. 2008. Advanced Reliability Models and Maintenance Policies. Springer, London.

Nakagawa, T. 2014. Random Maintenance Policies. Springer, London.

Papastavridis, S. and J. Hadjichristos. 1987. Mean time to failure for a consecutive-k-out-of-n:F system. IEEE Trans. Reliability R-36(1): 85-86.

Sasaki, M. et al. 1994. MTBF for consecutive-k-out-of-n:F systems with nonidentical component availabilities. IEICE Trans. Fundamentals E77-A(1): 122-128.

Tong, Y.L. 1985. A rearrangement inequality for the longest run, with an application to net-work reliability. Journal of Applied Probability 22: 386-393.

Zuo, M.J. and W. Kuo. 1990. Design and performance analysis of consecutive-k-out-of-n structure. Nav. Res. Logist. 37: 203-230.

Reliability Modeling and Estimation of (k_1, k_2)-out-of-(n, m) Pairs:G Balanced Systems

Elsayed A. Elsayed* and Jingbo Guo

Department of Industrial and Systems Engineering, Rutgers University, Piscataway, NJ 08854, USA

1. Introduction

Systems with units arranged in spatially distributed configurations are commonly found in many applications, such as power distribution and power generation systems. Unmanned Arial Vehicles (UAV) are another example of spatially distributed systems, with rotors distributed circularly in balanced configurations. UAVs have potential uses in military operations, commercial activities and other areas including aerial photography and surveying. However, poor controllability and reliability of the UAVs as well as the potential failures and damage to human lives and properties limit its applications. Hua and Elsayed (2016a,b) estimate the reliability metrics of k-out-of-n pairs:G balanced systems. More complex configurations of UAV systems are being developed where more than one level of rotors are arranged in parallel in order to improve their versatility and expand their applications. Reliability estimation and determination of mission completion of such UAVs are of interest to both the designers and operators of such systems.

In this chapter, we provide a detailed review of the current research on the reliability estimation of spatially distributed systems. We then address the multi-dimensional UAVs as (k_1, k_2)-out-of-(n, m) pairs:G

*Corresponding author: elsayed@soe.rutgers.edu

balanced systems. In such systems, n pairs of units are evenly distributed in a circular configuration. The angles between any two adjacent pairs are equal. In each side of a pair, there are m rotors arranged vertically in multilevel and in parallel with the other side of the pair. In other words, this is an n-pair-m-level UAV. At least k_2 out of m rotors have to operate properly on both sides of the pair for the pair to be considered as functional and balanced. The entire system is considered to be operating properly when at least k_1 out of the n pairs work properly and k_2 out of m rotors work properly in each operating pair while maintaining balance of the UAV. In this chapter, the balance of multilevel UAVs is discussed and presented as an alternative to the controllability methods used by UAV designers. Reliability estimation of such a system is obtained by enumerating the transition paths among all possible states of the system and by obtaining the probability of occurrence of all operational states.

This chapter is organized as follows: Section 2 provides a literature review of systems with spatially distributed units followed by reliability estimation of (k_1, k_2)-out-of-(n, m) pairs:G balanced systems in Section 3. Numerical examples are provided in section 4 in order to demonstrate the application of algorithms for reliability estimation. Summary of the chapter is provided in Section 5.

2. Literature review of systems with spatially distributed units

2.1 Multi-dimensional *k*-out-of-*n*:F systems

Multi-Dimensional k-out-of-n:F Systems are the most common systems with spatially distributed units. Unlike traditional k-out-of-n:F systems, which fail when k out of n units fail, in multi-dimensional k-out-of-n:F Systems, the number and the location of failed units together determine the state of the system. A similar system for multi-dimensional k-out-of-n: F systems is the consecutive k-out-of-n system, in which the number of consecutive failures and their locations are considered when estimating the system's reliability. The multi-dimensional k-out-of-n:F systems may have many different spatial configurations with different system requirements. The reliability estimation of these systems is, therefore, more difficult to obtain.

Salvia and Lasher (1990) introduce a two-dimensional consecutive k-out-of-n system. In their definition, the system is a square grid of size n and units are allocated at every intersection point. The system fails when a square grid of size k fails. Zuo (1993) extends the work by proposing a rectangular and cylindrical two-dimensional consecutive k-out-of-n system. Only bounds of the system failure probabilities are obtained.

Chang and Huang (2010) evaluate the reliability of the two-dimensional consecutive k-out-of-n system based on Markov Chain. More complex configurations are explored by other researchers. For example, Akiba and Yamamoto (2001) introduce the two-dimensional k-within-consecutive-$r \times s$-out-of-$m \times n$:F system. This system fails when k units in an $r \times s$-submatrix fail. They develop a recursive algorithm for the reliability estimation of the system. Yamamoto and Akiba (2005) also develop a new recursive algorithm for reliability estimation of a circularly connected-(r, s)-out-of-(m, n):F lattice system. In this system, $m \times n$ units are arranged in a cylindrical grid that contains m circles. Each circle has n units. When all units in a grid of size $r \times s$ fail, the system is considered failed. Similarly, an k-within-linear/circular connected-(r, s)-out-of-(m, n):F system is studied by Makri and Psillakis (1997). In these two systems, when k or more units fail in an $r \times s$-submatrix-out-of (m, n) matrix, the entire system is considered failed. Boehme et al. (1992) generalize the systems to a connected-X-out-of-(m, n):F lattice system. The reliability expressions for simple systems are provided. Godbole et al. (1998) expand the work into d-dimensional systems, which makes reliability estimation more challenging. Only the upper bound of reliability is obtained for these systems.

The three-dimensional consecutive k-out-of-n:F systems are also investigated by other researchers. Instead of modeling the system as a two-dimensional grid, Boushaba and Ghoraf (2002) propose a new system where if all units in a cubic grid of size k fail, the system is considered failed. They obtain the upper and lower reliability bounds for the system. Mahmoud Boushaba and Zineb Azouz (2011) introduce an intermediate system in order to estimate the lower bound of the three-dimensional system. Gharib et al. (2010) project the three-dimensional system to two or one dimensional system as special cases and study the reliability for three-dimensional k-within-$(2, 2, 2)$-out-of-$(m, 2, 2)$:F system. Gharid et al. (2011) later extend the three-dimensional cubic grid system to a $(1, 1, 2)$ or $(1, 2, 1)$ or $(2, 1, 1)$-out-of-$(n, 2, 2)$:F lattice system. This system is arranged as a cuboid of n layers. Each layer consists of 2×2 units. The system fails when at least two connected units fail. Akiba and Yamamoto (2004) study a more complicated 3-dimensional lattice system that fails when three units on at least one triangular fail.

Besides the many configurations of the systems presented above, other studies explore realistic situations. For example, Habib et al. (2010) consider the total number of failed units as a second condition for the system's failure. The system fails when the total number of failed units exceeds the maximum number allowed or when a consecutive (r, s)-out-of-(m, n) fails. This model is an extension of Yuge et al. (2003), in which a two-dimensional consecutive k-out-of-n:F system is investigated under the condition of restricted number of failed units.

2.2 *k*-out-of-*n* balanced system

In this chapter, we focus on multi-dimensional *k*-out-of-*n* model in order to study the reliability of Unmanned Aerial Vehicles since they are perforating into many applications such as military, aerospace, oceanography and others. However, the literature related to the reliability analysis for such systems is sparse. The difference between UAVs and other multi-dimensional *k*-out-of-*n* systems is that UAVs require system balance.

Sarper (2005) discusses the reliability estimation of Descent Systems of Planetary Vehicles for the future manned missions to Mars. This vehicle has four or six engines allocated evenly on a circle. Each engine along with its opposite engine form a pair that passes through the center of the circle, as shown in Fig. 1. To maintain balance, when one engine fails, the opposite engine in the same pair is forced down. He considers each pair as a correlated two-unit subsystem and uses bivariate exponential distribution to model the entire system. However, this approach is only applicable when the number of engines (and pairs) is small.

Hua and Elsayed (2016b) define balance of the system differently. In their system, multiple pairs of rotors (engines) are evenly distributed on a circular plane. Operational rotors are marked using 1 and failed rotors are marked using 0. The system is balanced when there exists at least one pair of perpendicular axes of symmetry. They introduce the concept of Moment Difference (MD) in order to determine the degree of symmetry with respect to any candidate axes that can be either along a pair or in the middle of two adjacent pairs. After examining the MD for every potential pair of perpendicular axes, if there are any pairs that have zero MD values for both axes, the system is considered balanced. As shown in Fig. 2, white circles and black circles represent operating and failed rotors, respectively. They are located evenly on a circular plane. Figure 2a is a balanced six-pair system since the system is symmetric with

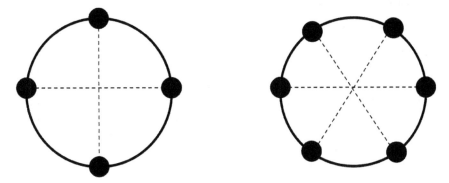

Fig. 1: Four and six engines descent system

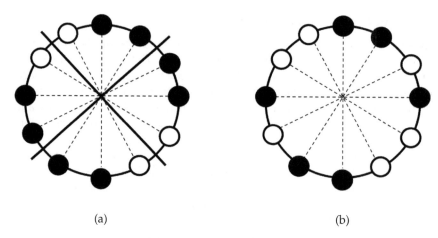

(a)   (b)

Fig. 2: Unbalanced and balanced six-pair system

respect to a pair of perpendicular axes (black solid lines). Without a pair of perpendicular axes of symmetry, the system in Fig. 2b is unbalanced and is considered failed. They define this model as k-out-of-n pairs:G balanced system. In this system, at least k-out-of-n pairs of the rotors need to operate properly for the UAV to provide its function while maintaining balance at all times. Not only are the reliability estimation and approximation methods obtained, the degradation analysis and load sharing effect are also considered (Hua and Elsayed 2016a, 2016b, Hua et al. 2015).

Another variation of the multi-dimensional k-out-of-n:F system are the recently designed rotary UAVs with several levels of rotors arranged in each pair. Modeling and reliability estimation of this system are introduced in the following section.

3. Reliability estimation for (k_1, k_2)-out-of-(n, m) Pairs:G balanced system

3.1 System description

We consider a rotary winged UAV with more than one level of rotors; several pairs of rotors are evenly arranged on a circular plane. The axis of each pair passes through the center of the plane and the angles between any adjacent pairs are equal. In each pair, there are more than one rotor arranged vertically in multi-level on each side. Figure 3 shows a 3-pair-2-level UAV designed by xFold Dragon (xFoldRig.com). We refer to such systems as (k_1, k_2)-out-of-(n, m) pairs:G balanced systems.

In such systems, n pairs of units are evenly distributed in a circular configuration. In each side of a pair, there are m rotors arranged vertically in multi-level and in parallel with the other side of the pair. In other words,

Fig. 3: Two-level UAV with three pairs of rotors ($n = 3$; $m = 2$)
(With permission from xFold Dragon)

this is an n-pair-m-level UAV. For each pair to be functioning, at least k_2 out of m rotors must operate properly on both sides of the pair, otherwise, the pair is considered failed and the rest of functioning rotors within this pair are forced down (standby mode). The system is considered to be operating properly when at least k_1 out of n pairs operate properly.

3.2 Balance requirement

To improve the reliability, the UAV designers use controllability methods such as control allocation when one or more rotors fail, as discussed in Achtelik et al. (2012). In this chapter, the concept of the operational balance is introduced as an alternative to the controllability methods. Based on the controllability testing method introduced by Du et al. (2015), for an PPNNPN hexacopter as shown in Fig. 4, the system is controllable when it is balanced and 2-out-of-3 pairs of rotors are operating properly ("P" denotes that a rotor rotates clockwise and is represented using white circles. "N" denotes that a rotor rotates anticlockwise and is represented by grey circles). This result shows that the concept of balance as well as the k-out-of-n requirement combined can be treated as an alternative approach in maintaining the controllability of UAVs.

We define the balance of (k_1, k_2)-out-of-(n, m) pairs:G balanced system as follows: Assume that the weights (or thrust forces) of all units are identical, the operating unit weighs 1 and the failed unit weighs 0. The system is considered balanced when its center of gravity is in the center of the circular plane.

In order to maintain balance, the opposite units of the failed ones are forced down as standbys. It is required that at least k_2 out of m units on

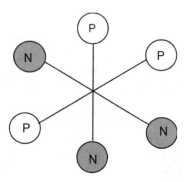

Fig. 4: PPNNPN hexacopter

both sides of a pair function for the pair to be functioning. At the same time, at least k_1 out of n pairs of units are required to function in order to meet the minimum requirement of a properly functioning system.

3.3 Reliability estimation for (k_1, k_2)-out-of-(n, m) pairs:G balanced system with identical units

Suppose all the units in the system are identical, then the system is modeled as a double loop k-out-of-n system. The outer loop k-out-of-n requires at least k_1-out-of-n pairs to be operating properly for the system to be functional. The inner loop k-out-of-n requires at least k_2-out-of-m units on both sides of a pair to be operating properly for the pair to be functioning.

Figure 5 shows a (1, 2)-out-of-(2, 4) system. In this system, two pairs of units (the first pair in black color and the second pair in grey color) are evenly distributed on a circle and at least one of the two pairs needs to be operational for the system to function properly. Each pair has four units ($m = 4$) allocated on both sides of the pair and two units need to be functioning properly for the pair to be operational. System balance requires that if one unit fails, then one of the operational units on the opposite side of the same pair must immediately be forced down as a standby. Since the failure of one unit causes the forcing down of another unit in the same pair, these two units can be considered as a two-unit series system and either of the units is chosen randomly to be forced down. In other words, each pair has m series sub-systems and k_2-out-of-m series sub-systems need to be operating for the pair to function properly. In this 2-pair-4-level system, each pair has four series sub-systems and two of them need to be operational. This is the inner loop k-out-of-n requirement of the system.

Suppose that the *pdf* and CDF of a single unit are f and F respectively. The *pdf* f_2 and CDF F_2 of a two-unit series sub-system are obtained as:

$$f_2(t) = 2 \times (1 - F(t)) \times f(t)$$

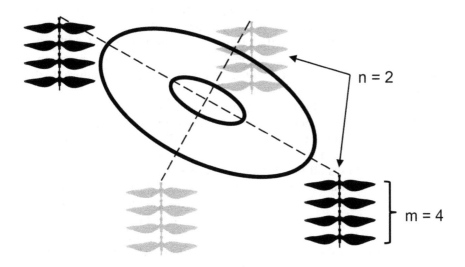

Fig. 5: $(1, 2)$-out-of-$(2, 4)$ pairs:G balanced system

$$F_2(t) = 1 - (1 - F(t))^2$$

The probability that a pair is operational at time t is:

$$G(t) = \sum_{l=k_2}^{m} \binom{m}{l} (1 - F_2(t))^l (F_2(t))^{m-l}$$

The reliability of the system at time t is:

$$R_s(t) = \sum_{j=k_1}^{n} \binom{n}{j} G(t)^j (1 - G(t))^{n-j}$$

3.4 Reliability estimation for (k_1, k_2)-out-of-(n, m) pairs:G balanced system with non-identical units

In this section, we estimate the reliability of (k_1, k_2)-out-of-(n, m) pairs:G balanced systems when the units are non-identical. It is obvious that this system is a redundant system and that the failure of units in certain positions and/or sequence may not cause the failure of the entire system. The operational states are the states under which the system is operational while maintaining balance. The reliability of the system at time t is the probability that the system is in one of the operational states at time t. Thus, in order to estimate the reliability of the system, one needs to obtain all the operational states and the probabilities that the system is in these states. This is achieved by keeping track of transition paths from one operational state to other possible operational states. Each transition between states is caused by the failure of one unit. The probability that the system is in a specific operational state at time t is the summation of probability that the system transits from the initial state (no failure) to this operational state during $(0, t)$.

We demonstrate this by using the $(1, 2)$-out-of-$(3, 3)$ system, shown in Fig. 6. In this system, there are three pairs of rotors (black, grey and light grey). There are three rotors allocated vertically on multi-level on both sides of each pair. Thus, this is a 3-pair-3-level system.

Figure 7 shows all the transitions between the operational states of this system. In the transition diagrams, vectors (l_1, l_2, l_3) represent the states of the system where l_i is the number of failed units in the ith pair. When one unit fails, one of the opposite units in the same pair is forced down immediately. Thus, l_i is also the number of forced down units in the ith pair. Since k_2-out-of-m units are required to function on both sides of a pair, when $l_i = k_2 + 1$, the remaining $m - (k_2 + 1)$ operating units on both sides are forced down since this pair is considered failed. In this system, since 2 out of 3 units are required to work on both sides of a pair, states such as $(3, 0, 0)$ or $(3, 1, 1)$ cannot exist. State $(2, 2, 2)$ is the failure state. All the transitions from the operational states eventually lead to this failure state. When two units fail or are forced down on both sides of a pair, the remaining units are forced down and the pair is considered failed. For example, the first pair in state $(2, 0, 0)$ has two units failed or forced down. Hence, this pair is considered failed and no further failures could occur. The transitions from this state to other states occur when there are new failures in the second or third pair. These transitions lead to states $(2, 0, 1)$ and $(2, 1, 0)$.

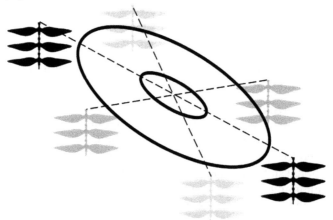

Fig. 6: $(1, 2)$-out-of-$(3, 3)$ pairs:G system

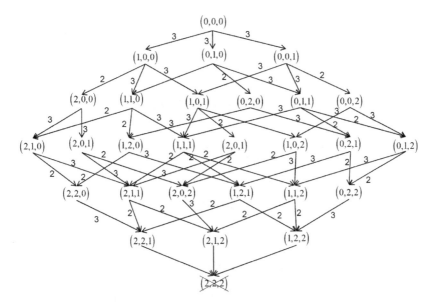

Fig. 7: Transition diagram of the (1,2)-out-of-(3, 3) pairs:G balanced system

After obtaining all the operational states based on the transition diagram, the probability that the system is in operational state (i, j, l) at time t is denoted as $P_{(i, j, l)}(t)$ (product of two parts: Number of possible transition paths from state $(0, 0, 0)$ to state (i, j, l) and the probability of transitioning from state $(0, 0, 0)$ to state (i, j, l). In Fig. 7, the number above each arrow is the number of transition paths from one operational state to another. The number of the transition paths from state $(0, 0, 0)$ to state (i, j, l) is the sum of the products of the number of transitions shown above the arrows from initial state $(0, 0, 0)$ to state (i, j, l). Taking states $(0, 0, 1)$ as examples, the probabilities that the system transits from state $(0, 0, 0)$ to one of the three states are equal. There are three possible transition paths from the initial state to the three states since the initial state transits to one of the three states immediately, as long as one of the m ($m = 3$) series subsystems fail. For state $(0, 1, 1)$, there are two different transitions from the initial state to the current state: $(0, 0, 0) \rightarrow (0, 1, 0) \rightarrow (0, 1, 1)$ or $(0, 0, 0) \rightarrow (0, 0, 1) \rightarrow (0, 1,1)$. The number of transition paths for both approaches is 3×3 = 9. The total number of transition paths from the initial state to state $(0, 1, 1)$ is 18. The probabilities for other operational states can be calculated similarly.

For simplicity, suppose that the units in same pairs are identical and that units in different pairs are non-identical. The *pdf* and CDF of a single unit for i^{th} pair are $f^i(t)$ and $F^i(t)$, $i = 1, ..., n$. The *pdf* $f_2^i(t)$, the CDF $F_2^i(t)$ and $F_2^i(t)$ are calculated, respectively, as follows:

$$f_2^i(t) = 2 \times (1 - F^i(t)) \times f^i(t)$$

$$F_2^i(t) = 1 - (1 - F^i(t))^2$$

$$R_2^i(t) = (1 - F^i(t))^2$$

The probability that the (1, 2)-out-of-(3, 3) pairs:G system is at states (1, 0, 0), (0, 1, 0) and (0, 0, 1) at time t is obtained, respectively, as:

$$P_{(1,0,0)}(t) = m \times \int_{\tau_1=0}^{t} f_2^1(\tau_1) d\tau_1 \times R_2^1(t)^{m-1} \times R_2^2(t)^m \times R_2^3(t)^m$$

$$= m \times F_2^1(\tau_1) \times R_2^1(t)^{m-1} \times R_2^2(t)^m \times R_2^3(t)^m$$

$$= 3 \times F_2^1(\tau_1) \times R_2^1(t)^2 \times R_2^2(t)^3 \times R_2^3(t)^3$$

$$P_{(0,1,0)}(t) = m \times \int_{\tau_1=0}^{t} f_2^2(\tau_1) d\tau_1 \times R_2^2(t)^{m-1} \times R_2^1(t)^m \times R_2^3(t)^m$$

$$= m \times F_2^2(\tau_1) \times R_2^2(t)^{m-1} \times R_2^1(t)^m \times R_2^3(t)^m$$

$$= 3 \times F_2^2(\tau_1) \times R_2^2(t)^2 \times R_2^1(t)^3 \times R_2^3(t)^3$$

$$P_{(0,0,1)}(t) = m \times \int_{\tau_1=0}^{t} f_2^3(\tau_1) d\tau_1 \times R_2^3(t)^{m-1} \times R_2^2(t)^m \times R_2^1(t)^m$$

$$= m \times F_2^3(\tau_1) \times R_2^3(t)^{m-1} \times R_2^2(t)^m \times R_2^1(t)^m$$

$$= 3 \times F_2^3(\tau_1) \times R_2^3(t)^2 \times R_2^2(t)^3 \times R_2^1(t)^3$$

Similarly, $P_{(0,1,1)}(t)$ is obtained as:

$$P_{(0,1,1)}(t) = m^2 \times \int_{\tau_2=0}^{t} \int_{\tau_1=0}^{\tau_2} f_2^2(\tau_1) f_2^3(\tau_2) d\tau_1 d\tau_2 \times R_2^1(t)^m \times R_2^2(t)^{m-1}$$

$$\times R_2^3(t)^{m-1} + m^2 \times \int_{\tau_2=0}^{t} \int_{\tau_1=0}^{\tau_2} f_2^3(\tau_1) f_2^2(\tau_2) d\tau_1 d\tau_2 \times R_2^1(t)^m$$

$$\times R_2^2(t)^{m-1} \times R_2^3(t)^{m-1}$$

$$= m^2 \times \int_{\tau_2=0}^{t} F_2^2(\tau_2) f_2^3(\tau_2) d\tau_2 \times R_2^1(t)^m \times R_2^2(t)^{m-1} \times R_2^3(t)^{m-1}$$

$$+ m^2 \times \int_{\tau_2=0}^{t} F_2^3(\tau_2) f_2^2(\tau_2) d\tau_2 \times R_2^1(t)^m \times R_2^2(t)^{m-1} \times R_2^3(t)^{m-1}$$

$$= \quad \times \int_{\tau_2=0}^{t} F_2^2(\tau_2) f_2^3(\tau_2) d\tau_2 \times R_2^1(t)^3 \times R_2^2(t)^2 \times R_2^3(t)^2$$

$$+ \quad \times \int_{\tau_2=0}^{t} F_2^3(\tau_2) f_2^2(\tau_2) d\tau_2 \times R_2^1(t)^3 \times R_2^2(t)^2 \times R_2^3(t)^2$$

The reliability of the system is the summation of these probabilities.

$$R(t) = P_{(0,0,0)}(t) + P_{(1,0,0)}(t) + \dots + P_{(1,2,2)}(t)$$

3.5 Algorithm for reliability estimation for (k_1, k_2)-out-of-(n, m) pairs:G balanced system

As shown in Section 3.4, estimating the reliability of (k_1, k_2)-out-of-(n, m) pairs:G balanced systems is time consuming even for systems with small values of n and m. For systems with more than 4 pairs and 2 levels of rotors, the enumeration of the operational is complicated. As mentioned, the greater the distance between the operational state and the initial state, the more transition paths there are. Each transition from one operational state to another requires an integral calculation in the reliability estimation expression. Thus, as the number of transitions increases, the integral becomes more complex. Therefore, we develop an algorithm to estimate the reliability for such systems.

The algorithm includes three parts: (1) Obtain all the operational states; (2) Obtain the number of transition paths for all the operational states; and (3) Calculate the probability that the system is in certain operational states at any time. They are described below.

Part 1: Obtain all operational states

Step 1: Calculate the maximum number of failed units in one pair: $l_{pair\_max} = m - k_2 + 1$.

Step 2: Calculate the maximum number of failed pairs in the system: $l_{sys\_max} = n - k_1$.

Step 3: Construct a vector to store all the new states and old state combination: C.

Step 4: Set an identity matrix of size $n \times n$: $\tilde{I}_n$ and a vector of ones of size $n \times 1$: $\tilde{1}_n$.

Step 5: Set the initial state (no failure) as a zero vector of size $1 \times n$: $S_{init} = \tilde{0}_{1 \times n}$. Place the state into a list: Q_{state}

Step 6: Take one state from Q_{state} and set it as an old state S_{old}. Delete the state from Q_{state}.

Step 7: Generate new states as follows: $S_{new} = \tilde{1}_n \times S_{old} + \tilde{I}_n$. Split the matrix S_{new} by rows and obtain all the new states: $S_{new}^1, S_{new}^2, ..., S_{new}^n$.

Step 8: Set a threshold value $h_i = 0$ for each new state $S_{new}^i, i = 1, ...n$. For every element in the vector $S_{new}^i(j), j = 1, 2, ..., n$, if $S_{new}^i(j) \geq l_{pair\_max}$, $h_i = h_i + 1$. If $h_i > l_{sys\_max}$, delete S_{new}^i, otherwise, place S_{new}^i in Q_{state} and store the combination of S_{new}^i and S_{old}^i in the list C.

Step 9: If Q_{state} is not empty, go back to step 6. If it is, stop the algorithm.

Part 2: Obtain the number of transition paths for all operational states

Step 1: Set a vector T of size $r \times 1$ to save the number of transition paths for each operational state: $T = 1_{r \times 1}$ where r is the total number of combinations in list $C_{r \times 2}$.

Step 2: For each new and old state combination C_i, $i = 1, 2, ..., r$, search for the duplicated combinations in list C. For each C_j, $j = i, (i+1), ...r$, if $C_j = C_i$, delete C_j and update T_i by adding 1. After finding all the combinations, T_i is the total number that the combination C_i exists in the original list C. After deleting all duplicates for each combination, the list C contains g unique combinations in total.

Step 3: For each combination C_i, $i = 1, 2 ..., g$, compare S^i_{old} and S^i_{new} and find the location where the numbers in the two vectors are different: d_i. Calculate the number of transitions from state S^i_{old} to state S^i_{new}: $(m - S^i_{old}(d_i))$. Update T_i by calculating the product of transition number and total number of C_i in list C: $T_i \times (m - S^i_{old}(d_i))$.

Part 3: Calculating the probability that the system is in certain operational states at any time

Step 1: Order list C based on the number of failed units in S_{new}. In combination C_1, S_{old} is a zero vector of size $1 \times n$.

Step 2: Define an expression for each new state in list C: p_i, $i = 1, ..., g$. $p_i = T_i \int_{\tau_1 = 0}^{t} p_{old}(\tau_1) f_2^{d_i}(\tau_1) d\tau_1$ where $p_{old}(\tau_1)$ is the expression of the old state in the ith combination. $p_{old}(\tau_1) = 1$ if the old state S_{old} is a zero vector of size $1 \times n$. $f_2^{d_i} = 2 \times (1 - F^{d_i}(t)) \times f^{d_i}(t)$, d_i is the location of difference for the ith combination and is obtained in part 2. Let $f^{d_i}(t)$ and $F^{d_i}(t)$ denote the *pdf* and CDF of a single unit in d_i pair.

Step 3: After obtaining the expressions p_i, $i = 1, ..., g$ for all the new states, let $R_2^{d_i}(t) = (1 - F^{d_i}(t))^2$. $p_i(t) = p_i(t) \times \prod_{d_i = 1}^{n} R_2^{d_i}(t)^{l_{d_i}}$ where $l_{d_i} = m - S^i_{new}(d_i)$.

Step 4: $p_i(t)$, $i = 1, 2, ... g$ is the probability that the system is in new state S^i_{new} at any time t. Reliability of the system is the summation of these probabilities, $R(t) = \sum_{i=1}^{g} p_i(t)$.

The flow chart for this algorithm is shown in Fig. 8.

4. Numerical examples

4.1 Numerical examples for systems with identical units

Scenario 1: Reliability estimation for systems with identical units
In this section, a numerical example is provided in order to illustrate the reliability estimation for the (k_1, k_2)-out-of-(n, m) pairs:G balanced system

when units are identical. Suppose that all units in this 2-pair-4-level system follow an exponential distribution with failure rate λ. The expressions for this system are obtained as follows:

$$f_2(t) = 2 \times (1 - F(t)) \times f(t) = 2 \times \exp^{-\lambda t} \times \lambda \exp^{-\lambda t} = 2\lambda \exp^{-2\lambda t}$$

$$F_2(t) = 1 - (1 - F(t))^2 = 1 - (1 - (1 - \exp^{-\lambda t}))^2 = 1 - \exp^{-2\lambda t}$$

The reliability functions are shown in Fig. 9. Units in the system follow exponential distribution with three different failure rates: $\dfrac{1}{90}, \dfrac{1}{60}, \dfrac{1}{30}$.

Scenario 2: Reliability estimation for systems with different k_1, k_2, n and m

Reliability estimations for the (k_1, k_2)-out-of-(n, m) pairs:G balanced systems with different combinations of k_1, k_2, n and m are calculated when units are identical. It shows that increasing k_1 and k_2, and decreasing n and m leads to a decrease in the system reliability.

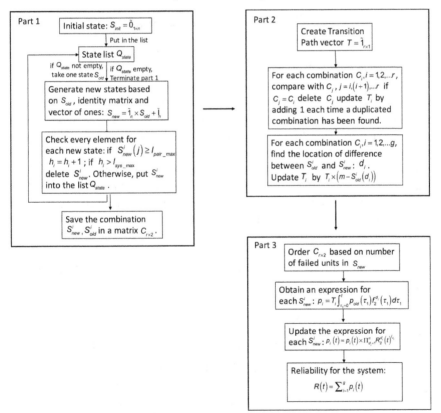

Fig. 8: Flow chart of the algorithm

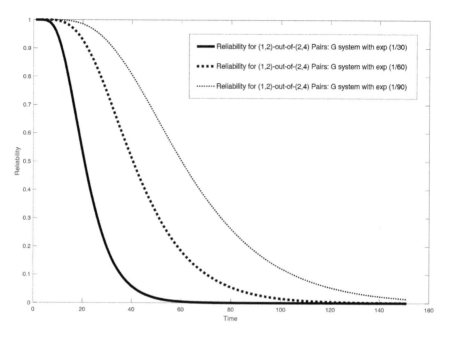

Fig. 9: Reliability for (k_1, k_2)-out-of-(n, m) pairs:G balanced system

Scenario 3: System design with fixed $k_1 \times k_2$ and $n \times m$

Suppose a system has a requirement for the minimum number of operating units (thrust forces), thus, the product $k_1 \times k_2$ is fixed. Also, suppose that the number of available units is also limited, i.e. $n \times m$ is fixed. We are interested in designing a system under these assumptions that results in the arrangement with highest reliability. Figure 11 shows the reliability plots for systems with 16 available units and the minimum number of operating units is four when all units follow exponential distribution with mean life time equal to 30. From the figure $(2, 1)$-out-of-$(4, 2)$ pairs:G balanced system has the highest reliability; $(1, 2)$-out-of-$(4, 2)$ pairs:G balanced system has the lowest reliability. The system tends to have higher reliability when both the ratios $\dfrac{k_1}{n}$ and $\dfrac{k_2}{m}$ are close to $\dfrac{1}{2}$. A theoretical proof will be investigated in future research.

4.2 Numerical examples for systems with non-identical units

In this section, an example is introduced in order to illustrate the reliability estimation of (k_1, k_2)-out-of-(n, m) pairs:G balanced system when the units are non-identical. For a $(1, 2)$-out-of-$(3, 3)$ pairs:G balanced system, there are three pairs located evenly on a circular plane. In each pair, there are three levels. In this system, suppose that the units in the same pair

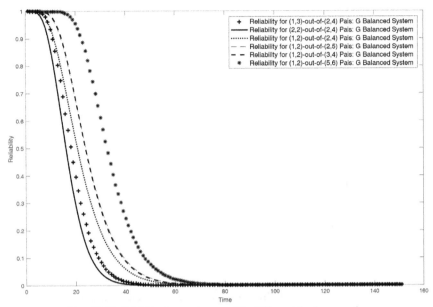

Fig. 10: Reliability for systems with different k_1, k_2, n and m

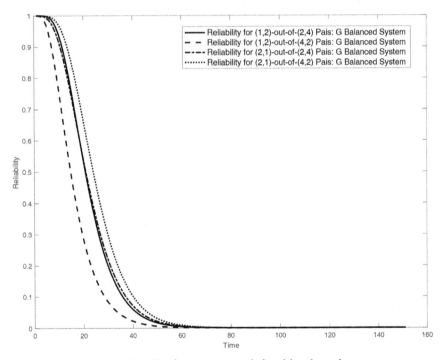

Fig. 11: Reliability for systems with fixed $k_1 \times k_2$ and $n \times m$

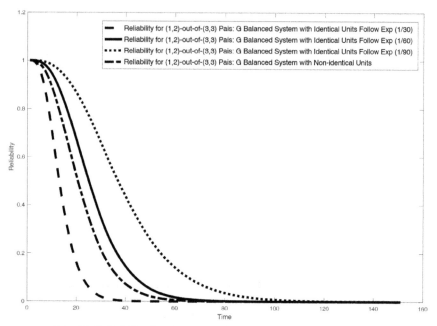

Fig. 12: Reliability for (1,2)-out-of-(3,3) pairs:G balanced system

follow the same distribution, and units in different pairs follow different distributions. Suppose units in the ith pair follow exponentially with failure rate: λ_i.

Suppose $\lambda_1 = \dfrac{1}{30}, \lambda_2 = \dfrac{1}{60}, \lambda_3 = \dfrac{1}{90}$, reliability functions for (1, 2)-out-of-(3, 3) pairs:G balanced system when all units are identical and follow exponential distribution with failure rates $\lambda : \dfrac{1}{90}, \dfrac{1}{60}, \dfrac{1}{30}$, respectively, as shown in Fig. 12. The reliability of the system with non-identical units is higher than the system with identical units when the failure rate is highest: $\dfrac{1}{30}$. However, this reliability curve is lower than the other two systems when units are identical, and the failure rates are $\dfrac{1}{90}, \dfrac{1}{60}$, respectively. The reason for this is that the pair with the highest failure rate of $\dfrac{1}{30}$ fails rapidly and influences the reliability for the entire system.

5. Summary

In this chapter, we present a detailed review of the current research on the reliability estimation of spatially distributed systems. Different spatially distributed systems, including one level and multi-level Unmanned Ariel

Vehicles, are discussed. The operational balance requirements for these systems are considered as an alternative to control and dynamic theory when one or more rotors fail in UAVs. In order to obtain the reliability, such systems are modeled as (k_1, k_2)-out-of-(n, m) pairs:G balanced system. The challenges and difficulties of reliability estimation for such systems include the enumeration of transitions between states and the multiple integral calculations. Further work is needed in order to investigate the reliability for the systems when forced down units can resume operation while meeting the minimum requirements of operating pairs and the system's balance requirement.

REFERENCES

Achtelik, M., K.-M. Doth, D. Gurdan and J. Stumpf. 2012. Design of a multi rotor MAV with regard to efficiency, dynamics and redundancy. *In*: AIAA Guidance, Navigation, and Control Conference, 13-16 August 2012 Minneapolis, USA.

Akiba, T. and H. Yamamoto. 2001. Reliability of a 2-dimensional k-within-consecutive-r' s-out-of-m' n:F system. Naval Research Logistics (NRL) 48: 625–637.

Akiba, T. and H. Yamamoto. 2004. Upper and lower bounds for 3-dimensional k-within-consecutive-(r_1, r_2, r_3)-out-of-(n_1, n_2, n_3):F system. *In*: Proceedings of the 2004 Asian International Workshop, 26–27 August 2004 Hiroshima, Japan.

Boehme, T.K., A. Kossow and W. Preuss. 1992. A generalization of consecutive-k-out-of-n:F systems. IEEE Transactions on Reliability 41(3): 451–457.

Boushaba, M. and N. Ghoraf. 2002. A 3-dimensional consecutive-k-out-of-n:F models. International Journal of Reliability, Quality & Safety Engineering 9(2): 193–198.

Chang, Y.M. and T.H. Huang. 2010. Reliability of a 2-dimensional k-within-consecutive-r, s-out-of-m, n:F system using finite Markov chains. IEEE Transactions on Reliability 59(4): 725–733.

Du, G.-X., Q. Quan and K.-Y. Cai. 2015. Controllability analysis and degraded control for a class of hexacopters subject to rotor failures. Journal of Intelligent & Robotic Systems 78(1): 143–157.

Gharid, M., E.M. Elsayed and I.I.H. Nashwan. 2010. Reliability of simple 3-dimensional consecutive-k-out-of-n:F systems. Journal of Mathematics and Statistics 6(3): 261–264.

Gharid, M., E.M. Elsayed and I.I.H. Nashwan. 2011. Reliability of connected (1,1,2) or (1,2,1) or (2,1,1)-out-of-$(n,2,2)$:F lattice systems. Journal of Advanced Research in Statistics and Probability 3(1): 47–56.

Godbole, A.P., L.K. Potter and J.K. Sklar. 1998. Improved upper bounds for the reliability of d-dimensional consecutive-k-out-of-n:F systems. Naval Research Logistics (NRL), 45: 219–230.

Habib, A.S., T. Yuge, R.O. Al-Seedy and S. Ammar. 2010. Reliability of a consecutive-*r'* *s*-out-of-*m'* *n*:F lattice system with conditions on the number of failed components in the system. Applied Mathematical Modeling, 34(3): 531–538.

Hua, D. and E. Elsayed. 2016a. Degradation analysis of *k*-out-of-*n* pairs:G balanced system with spatially distributed units. IEEE Transactions on Reliability 65(2): 941–956.

Hua, D. and E.A. Elsayed. 2016b. Reliability estimation of *k*-out-of-*n* pairs:G balanced systems with spatially distributed units. IEEE Transactions on Reliability 65(2): 886–900.

Hua, D., E.A. Elsayed, K.N. Al-Khalifa and A.S. Hamouda. 2015. Reliability estimation of load sharing capacity-*c*-out-of-*n* pairs:G balanced system. *In:* Prognostics & System Health Management Conference (PHM), 21–23 October 2015 Beijing, China.

Mahmoud Boushaba and Zineb Azouz. 2011. Reliability bounds of a 3-dimensional consecutive-*k*-out-of-*n*:F system. International Journal of Reliability, Quality and Safety Engineering 18(1): 51–59.

Makri, F.S. and Z.M. Psillakis. 1997. Bounds for reliability of *k*-within connected-*r'* *s*-out-of-*m'* *n* failure systems. Microelectronics Reliability 37: 1217–1224.

Salvia, A.A. and W.C. Lasher. 1990. Two-dimensional consecutive-*k*-out-of-*n*:F models. IEEE Transactions on Reliability 39: 382.

Sarper, H. 2005. Reliability analysis of descent systems of planetary vehicles using bivariate exponential distribution. *In:* Reliability and Maintainability Symposium Proceedings, 24-27 January 2005 Alexandria, USA.

Yamamoto, H. and T. Akiba. 2005. A recursive algorithm for the reliability of a circular connected-(*r*, *s*)-out-of-(*m*, *n*): F lattice system. Computers & Industrial Engineering 49: 21–34.

Yuge, T., M. Deher and S. Yanagi. 2003. Reliability of a 2-dimensional consecutive-*k*-out-of-*n*:F system with a restriction in the number of failed components. IEICE Transactions on Fundamentals of Electronics Communications and Computer Sciences, E86A(6): 1535–1540.

Zuo, M. 1993. Reliability and design of 2-dimensional consecutive-*k*-out-of-*n* systems. IEEE Transactions on Reliability 42(3): 488–490.

Sequencing Optimization for *k*-out-of-*n*:G Cold-standby Systems Considering Reliability and Energy Consumption

Prashanthi Boddu[1], Liudong Xing[2*] and Gregory Levitin[3]

[1] Global Prior Art Inc, Boston, MA, USA
[2] Electrical and Computer Engineering Department, University of
  Massachusetts, Dartmouth, USA
[3] The Israel Electric Corporation, Haifa 31000, Israel

1. Introduction

Traditional system reliability optimization problems, such as the redundancy allocation problem (RAP), are proven to be NP-hard (Chern 1992). Many solution methodologies have been proposed to solve them (Gen and Yun 2006, Kuo et al. 2001, Kuo and Wan 2007). Research findings on solving the traditional reliability optimization problems mostly pertain to series-parallel systems with either 1-out-of-*n*:G or *k*-out-of-*n*:G sub-system configuration where at least 1 or *k* components should be working for the sub-system to function, respectively. In addition, each parallel/ standby sub-system has been limited to either hot or cold standby redundancy. A hot standby unit operates in synchrony with the online primary unit and provides immediate replacement once the online unit fails; while a cold-standby unit is initially unpowered and is switched into the power mode when it is needed to replace the failed online unit. Therefore, the hot standby redundancy is generally used for applications

*Corresponding author: lxing@umassd.edu

where the recovery time is critical; the cold standby redundancy is commonly used for applications where the energy consumption is critical.

Specifically, exact optimization methods, like dynamic programming (Fyffe et al. 1968), Lagrangean multipliers (Misra 1972), and integer programming (Misra 1991), have been proposed for solving the RAP of 1-out-of-n:G series-parallel systems with hot redundancy and homogeneous backup scheme where one type of components can be substituted only by the same type of components to achieve fault tolerance. Later on, meta-heuristic optimization methods such as Genetic Algorithm, Ant Colony Optimization Algorithm, the improved surrogate constraint method, and Tabu search (Chen and You 2005, Coit and Smith 1996, Liang and Smith 2004, Onishi et al. 2007) were proposed to solve the RAP of either 1-out-of-n:G or k-out-of-n:G series-parallel systems with hot standby redundancy and adapted heterogeneous backup scheme where one type of components can be substituted with a different type of functionally equivalent components to achieve fault tolerance. In (Coit 2001), a solution methodology based on integer programming was proposed in order to determine the optimal design configuration for non-repairable 1-out-of-n:G series-parallel systems with cold standby redundancy. In (Coit 2008, Coit and Liu 2000, Tavakkoli-Moghaddam et al. 2008), a solution methodology was proposed for solving the RAP of 1-out-of-n:G or k-out-of-n:G heterogeneous series-parallel systems where standby sub-systems exclusively involve either hot or cold standby redundancy. None of the approaches mentioned above consider the combination of hot and cold standby redundancy techniques within the same standby sub-system, though such combination can achieve a balance between fast recovery and energy conservation. The mixed redundancy technique can be useful for practical optimal system designs in applications such as space exploration and satellite systems, aerospace systems, defense systems, textile manufacturing systems, carbon recovery systems used in fertilizer plants, etc. (Coit 2001). Recently, Boddu et al. have solved the RAP for 1-out-of-n:G and k-out-of-n:G heterogeneous series-parallel systems with the mixed hot and cold standby redundancy for each standby sub-system (Boddu et al. 2011, Boddu and Xing 2011, 2012).

Recent works on standby sparing systems also revealed that the component activation sequence can greatly affect the performance of standby systems with heterogeneous types of components, leading to a new type of optimization problems called the optimal standby element sequencing problem (SESP) (Levitin et al. 2013a). The objective of SESP is to select an initiation sequence of system components maximizing the system reliability, or minimizing the expected system cost while meeting a desired level of system reliability. Based on an event transition-based numerical methodology, the SESP problem has been formulated and solved for different types of standby systems, including cold standby systems (Levitin et al. 2013a, b), mixed hot and cold standby systems (Levitin et al. 2014a, 2015a), warm standby systems (Levitin et al. 2013c, 2014b), and systems with standby mode transfers (Dai et al. 2017, Levitin et al. 2014c, 2015b, c). Furthermore, joint optimal SESP and checkpoint distribution problems have been solved for standby computing systems with different checkpointing schemes assisting efficient system recovery in the case of a component failure occurring (Levitin et al. 2015, 2015d, 2015e, 2016a, 2016b, 2017, 2018).

To the best of our knowledge, little work has considered the energy consumption explicitly despite it being a critical performance and design factor for standby systems. For example, in (Cao 2012) the RAP was solved for standby systems subject to cost and energy consumption constraints, where the energy consumption of the system is simply modeled as the sum of energy consumption of all the selected component choices for the system design without considering the components' real life time. Such a model is not accurate as the energy consumption should be a function of actual operation time of the system components. In standby systems, the operational time of a component is dynamic and random. Thus, the total energy consumption of the standby systems exhibits random behavior given the constant power consumption rates of the components. In (Boddu et al. 2014), an analytical model was suggested for analyzing the practical energy consumption of 1-out-of-n:G cold-standby systems.

In this chapter, we extend the work of (Boddu et al. 2014) for the energy consumption analysis of k-out-of-n:G cold-standby systems. Based on the extended energy consumption model, the SESP is solved with the objective of minimizing the expected system energy consumption while satisfying a constraint on the overall system reliability.

The remainder of the chapter is organized as follows: Section 2 presents the problem description and formulation. Section 3 presents an analytical method for the energy consumption and reliability analysis of a k-out-of-n:G cold-standby system, followed by example analyses using the proposed method. Section 4 presents the optimal sequencing results for several cases using the brute force approach. Section 5 presents conclusions and directions for future work.

2. Problem description and formulation

This section presents the problem description and formulation of the optimal SESP for a k-out-of-n:G heterogeneous cold-standby system, as shown in Fig. 1.

2.1 Problem description

Figure 1(a) represents the dynamic fault tree (DFT) model of a k-out-of-n:G cold-standby system using the Cold SPare (CSP) gate (Xing and Amari 2008). Figure 1(b) illustrates the corresponding block diagram model. The cold-standby system has k primary units ($P_1, P_2, ..., P_k$) and nc cold-standby units. The predetermined total number of system components represented as n is given by $nc+k$. For a k-out-of-n:G cold standby system, it is assumed that initially k primary components are online and operating. When any one of the k online/primary components fails, it is replaced with the first cold-standby unit in the prescribed sequence. Then, when any one of the k operating units fails, it is replaced with the next cold-standby unit in the sequence. The cold-standby system fails when $nc+1$ units have failed.

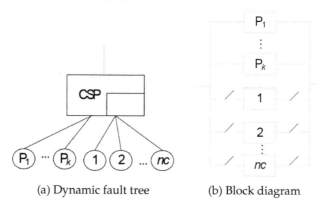

(a) Dynamic fault tree   (b) Block diagram

Fig. 1: Models of k-out-of-n:G cold-standby system

For the cold-standby system, there is a single pool of multiple component choices that are functionally equivalent. Each component choice is characterized by a specific level of constant power consumption rate, constant start-up energy consumption and reliability. It is assumed that the cold-standby system has perfect fault detection and switching mechanisms. The set of components is given. Both reliability and energy consumption of a k-out-of-n:G cold-standby system depends on the selection of k primary and nc cold standby components as well as on the sequence of putting the selected cold-standby units into operation when the desired level of system reliability is high. Note that when the mission time is large (system reliability is very low), the energy consumption of the cold-standby system with different initiation sequences converges to the same value since all the cold standby components tend to be activated and then fail. Therefore, the objective of the optimization problem in this work is to select the k primary units and nc cold-standby units from the given discrete component choices and to select the sequence of initializing/ activating cold-standby units so as to minimize the expected total system energy consumption while meeting the system-level constraint on the overall system reliability.

2.2 Problem formulation

Equation (1) defines the optimization problem considered in this work:

$$\text{minimize } \overline{E(t)} \text{ subject to } R(t) \geq R^* \tag{1}$$

where $\overline{E(t)}$ is the expected total energy consumption of the system during its mission time t, which is modeled as a non-linear function, depending on the sequence of putting the cold-standby units into operation. An analytical model is proposed in the next section for the analysis of this expected value. $R(t)$ in (1) is the overall system reliability, i.e., the probability that no more than nc elements in the sequence fail before mission time t. R^* is the desired level of system reliability.

3. Proposed energy consumption model

This section describes an analytical model for the energy consumption analysis and reliability analysis of the k-out-of-n:G cold-standby system in Fig. 1, followed by the analysis of example systems using the proposed method. The following notations are used:

V	vector representing the sequence of elements in the system. The selected k primary and nc cold-standby versions are positioned respectively from $(1, 2, ..., k)$ and $(k + 1, ..., n)$.
FE_i	event that exactly i elements fail before time t and the remaining elements in the system fail after time t for any $i = 0, 1, ..., n$.
f_i	total number of valid failure sequences/permutations for event FE_i
$FE_{i,p}$	pth failure sequence event/vector for FE_i
$\Pr(FE_{i,p})$	occurrence probability of $FE_{i,p}$
$E_{i,p}(t)$	energy consumption of the cold-standby system given the occurrence of $FE_{i,p}$, $i = 0, 1, ..., n$
$\overline{E_{i,p}(t)}$	mean of $E_{i,p}(t)$
$FE_{i,p}(z)$	type of primary/cold-standby unit in vector $FE_{i,p}$ located at position z.
$W_{i,ph}/W_{i,pc}$	set of primary units/powered cold-standby units that are still working in the system for the pth failure sequence $FE_{i,p}$ during the mission time.

$F_{i,ph}/F_{i,pc}$	set of primary units/powered cold-standby units that have failed in the system for the pth failure sequence $FE_{i,p}$ during the mission time.
T_z	r.v. representing the life time of a primary/cold-standby unit in $V/FE_{i,p}$ located at position z.
$\mu[T_z]$	mean of T_z
$f_{T_z}(t)$	probability density function (*pdf*) of r.v. T_z.
$r_{FE_{i,p}(z)}$	power rate of the unit of type $FE_{i,p}$ (z)
$\lambda_{FE_{i,p}(z)}$	failure rate of the unit of type $FE_{i,p}$ (z)
$e_{oFE_{i,p}(z)}$	startup energy consumption of the unit of type $FE_{i,p}$ (z)
t	mission time
$Q(t)$	system unreliability at t
$1(A)$	unity function: $1(A) = 1$ if A is true, $1(A) = 0$ if A is false

3.1 Energy consumption model

The method of evaluating $\overline{E(t)}$ in (1) can be described as the following step-by-step procedure.

Step 1: Construct a failure event (*FE*) space for the k-out-of-n:G cold standby system where each *FE* denoted by FE_i represents the event that exactly i elements fail before time t and the remaining elements in the system fail after time t for any $i = 0, 1, ..., n$. It is assumed that even when the system fails after $nc + 1$ components have failed before time t, the remaining components can continue to operate and consume energy until time t. On the other hand, if the entire system is shut down whenever it fails, the remaining components will stop operating and consuming energy, we only need to consider the first $(nc + 2)$ failure events, i.e., FE_i for $i = 0, 1, ..., nc + 1$.

Establish the valid sequences of random unit failure times represented as $FE_{i,p}$ for the corresponding event FE_i where $p = 1, 2, ...f_i$. A failure sequence is invalid if the order in which a cold-standby unit fails is invalid. The k primary online units operate from time zero. All cold units operate from the time when they are powered on to replace the faulty online unit until they fail. Let T_z represent the life time of the unit in the position z of the failure sequence $FE_{i,p}$. Thus, it is the time-to-failure of the primary unit with starting time from time 0 or the time-to-failure of the powered cold-standby unit with starting time being the failure time of the previous unit in the sequence. For a particular failure sequence $FE_{i,p}$, sets $F_{i,ph}/F_{i,pc}$ and $W_{i,ph}/W_{i,pc}$ can be extracted, respectively, from i failed and the remaining $(n-i)$ working primary/powered cold-standby units. The occurrence probability of each failure sequence $FE_{i,p}$ can be evaluated using (2):

$$\Pr(FE_{i,p}) = B \int_0^t \int_{t_1}^t .. \int_{t_{i-1}}^t ACdt_i....dt_2dt_1,$$

where

$$A = \prod_{z=1}^i \left[f_{T_z}(t_z)^{1\left(FE_{i,p}(z)\in F_{i,ph}\right)} f_{T_z}(t_z - t_{c(z)})^{1\left(FE_{i,p}(z)\in F_{i,pc}\right)} \right]$$

$$B = \prod_{z=1}^k (1 - \int_0^t f_{T_z}(t_z)dt_z)^{1\left(V(z)\in W_{i,ph}\right)} \qquad (2)$$

$$C = \prod_{z=k+1}^n (1 - \int_{t_{z-k}}^t f_{T_z}(t_z - t_{z-k})dt_z)^{1\left(V(z)\in W_{i,pc}\right)}$$

In (2), the term A is the product of *pdf* of *r.v* representing the life time of primary/cold-standby units that have failed in the particular failure sequence $FE_{i,p}$; $t_{c(z)}$ is the time at which the cold-standby unit is powered on, where $c(z)$ = [(position of cold-standby unit of type $FE_{i,p}(z)$ in vector V)–k]. The term B is the product of the reliability functions of all hot units that have survived the mission time for the particular failure sequence. The term C is the product of the reliability functions of all the powered cold-standby units that have survived the mission time for the particular failure sequence. The summation of occurrence probabilities of all the FE is ONE, i.e., $\sum_{i=0}^n \sum_{p=1}^{f_i} Pr\{FE_{i,p}\} = 1$. The summation of all the failure sequences of the events where at least $n-k+1$ units fail before t, i.e. $\sum_{i=n-k+1}^n \sum_{p=1}^{f_i} Pr\{FE_{i,p}\}$, actually gives the overall system unreliability $Q(t)$. The system reliability $R(t) = 1 - Q(t)$.

Step 2: Develop the energy consumption equation for each failure sequence $FE_{i,p}$, where $i = 0, 1, ..., n$ and $p = 1, 2, ... f_i$. Energy consumption is the product of power rate of the component and its actual operational time or life time. Given the start-up energy consumption and power rate of each unit, the energy consumption of the cold-standby system given the occurrence of the failure sequence $FE_{i,p}$, denoted by $E_{i,p}(t)$, where $i = 0, 1, ..., n$ and $p = 1, 2, ..., f_i$ can be evaluated using (3).

$$E_{i,p}(t) = X + Y + Z, \text{ where}$$

$$X = \sum_{z=1}^i [r_{FE_{i,p}(z)} * T_z * 1\left(FE_{i,p}(z) \in F_{i,ph}\right)$$

$$+ r_{FE_{i,p}(z)} * T_z * 1\left(FE_{i,p}(z) \in F_{i,pc}\right)] + \sum_{z=1}^i e_{0FE_{i,p}(z)}$$

$$Y = \sum_{z=1}^{k} (r_{V(z)} * t + e_{0V(z)}) * 1(V(z) \in W_{i,ph})$$

$$Z = \sum_{z=k+1}^{n} (r_{V(z)} * (t - T_{s,z}) + e_{0V(z)}) * 1(V(z) \in W_{i,pc}) \tag{3}$$

In (3), the term X represents the energy consumption of all the failed primary/powered cold-standby units. The term Y represents the energy consumption of all the primary units that have survived the mission time t. The term Z represents the energy consumption of all the powered cold-standby units that have survived the mission time, where $T_{s,z}$ is the time at which the cold-standby unit is powered on.

Step 3: Evaluate the expected system energy consumption $\overline{E(t)}$ for the cold-standby system.

For evaluating $\overline{E(t)}$, we calculate the mean of each $E_{i,p}(t)$, denoted by $\overline{E_{i,p}}(t)$ with respect to random variables T_z:

$$E_{i,p}(t) = \mu(X + Y + Z), \text{ where}$$

$$\mu(X) = \sum_{z=1}^{i} [r_{FE_{i,p}(z)} * \mu(T_z) * 1\left(FE_{i,p}(z) \in F_{i,ph}\right)$$

$$+ r_{FE_{i,p}(z)} * \mu(T_z) * 1\left(FE_{i,p}(z) \in F_{i,pc}\right)] + \sum_{z=1}^{i} e_{0FE_{i,p}(z)} \tag{4}$$

$$\mu(Y) = \sum_{z=1}^{k} (r_{V(z)} * t + e_{0V(z)}) * 1(V(z) \in W_{i,ph})$$

$$\mu(Z) = \sum_{z=k+1}^{n} (r_{V(z)} * (t - \mu(T_{s,z})) + e_{0V(z)}) * 1(V(z) \in W_{i,pc})$$

In general, $\overline{E_{i,p}}(t)$ can be simplified and represented as:

$$\overline{E_{i,p}}(t) = \sum_{z=1}^{i} w_z \mu(T_z) + E_c \tag{5}$$

where w_z is a combined power consumption rate, defined as

$$w_z = \begin{cases} \sum_{z=1}^{i} [r_{FE_{i,p}(z)} * 1\left(FE_{i,p}(z) \in F_{i,ph}\right) + r_{FE_{i,p}(z)} * 1\left(FE_{i,p}(z) \in F_{i,pc}\right)] \\[2ex] - \sum_{z=k+1}^{n} (r_{V(z)} * 1(V(z) \in W_{i,pc}) \text{ if } (T_{s,z} \text{ in term } Z) = (T_z \text{ in term } X) \\[2ex] \sum_{z=1}^{i} [r_{FE_{i,p}(z)} * 1\left(FE_{i,p}(z) \in F_{i,ph}\right) + r_{FE_{i,p}(z)} * 1\left(FE_{i,p}(z) \in F_{i,pc}\right)] \text{ otherwise} \end{cases} \tag{6}$$

E_c in (5) is a constant energy consumption value for the given mission time t, defined as

$$E_c = \sum_{z=1}^{i} e_{0FE_{i,p}(z)} + \sum_{z=1}^{k} \left((r_{V(z)} * t + e_{0V(z)}) * 1(V(z) \in W_{i,ph}) \right)$$

$$+ \sum_{z=k+1}^{n} \left((r_{V(z)} * t + e_{0V(z)}) * 1(V(z) \in W_{i,pc}) \right)$$

(7)

According to the total probability law, the expected total energy consumption of the cold-standby system is

$$\overline{E(t)} = \sum_{i=0}^{n} \sum_{p=1}^{f_i} \Pr\{FE_{i,p}\} * \overline{E_{i,p}(t)}$$

(8)

In this method, the life time of each unit T_z can follow any arbitrary distribution. In the case of T_z following the exponential distribution with pdf $f_{Tz}(t) = \lambda_{FE_{i,p}(z)} e^{-\lambda_{FE_{i,p}(z)} t}$ and mean of $1/\lambda_{FE_{i,p}(z)}$, based on (5), $\overline{E_{i,p}(t)}$ in (8) can be calculated as:

$$\overline{E_{i,p}(t)} = \sum_{z=1}^{i} w_z \mu(T_z) + E_c$$

$$= \sum_{z=1}^{i} w_z / \lambda_{FE_{i,p}(z)} + E_c$$

(9)

3.2 Illustrative examples

In this section, we analyze the expected total energy consumption and reliability of two example cold-standby systems with 2-out-of-3:G and 2-out-of-4:G configuration, respectively.

Example 1: 2-out-of-3:G cold-standby system ($n = 3$ and $k = 2$)

Assume the given set of three components contains components of types 1, 2 and 3, respectively. They have non-identical power rates of $r_1 = 4$ W, $r_2 = 3$ W, $r_3 = 4.5$ W, failure rates of $\lambda_1 = 0.001/$hr, $\lambda_2 = 0.002/$hr, $\lambda_3 = 0.003/$hr, and start-up energy consumption of $e_{01} = 5$ Wh, $e_{02} = 4$ Wh, $e_{03} = 6$ Wh. With the given set of components, there are totally three possible 2-out-of-3 cold-standby system designs: [1 2 3], [1 3 2], and [3 2 1]. The system fails when $n-k+1$ (=2) or more components fail. In the following we evaluate $\overline{E(t)}$ for each of the system designs with detailed illustration for $V = [1\ 2\ 3]$. Assume mission time $t = 200$ hrs, and $R^* = 0.7$.

Solution for $V = [1\ 2\ 3]$: Meaning the system has two primary units of type 1 and type 2 and one cold-standby unit of type 3. Initially the primary units of type 1 and type 2 are online. When any one of the primary unit (say type 2) has failed, it is replaced with the cold unit of type 3 in the sequence. The system then fails when either primary unit of type 1 or powered cold unit of type 3 has failed.

Step 1: For the given system design, four failure events FE_i, where $i = 0$, 1, 2, 3 are possible. Next we describe the failure sequences corresponding to each event FE_i and the occurrence probabilities of $FE_{i,p}$ where $p = 1$, 2, ... f_i.

(a) Event FE_0: No unit fails before t. There is only one possible combination ($f_0 = 1$), i.e., all $k = 2$ primary components survive the mission time. In particular,

$FE_{0,1} = [1\ 2]$: the primary units of type 1 and type 2 survive mission time t.

$$\Pr\{FE_{0,1}\} = \left(1 - \int_0^t f_{T_1}(t_1)dt_1\right) * \left(1 - \int_0^t f_{T_2}(t_2)dt_2\right) = 0.5488$$

(b) Event FE_1: One unit fails before t. There are two possible failure sequences ($f_1 = 2$): $FE_{1,1} = [1\ |\ 2\ 3]$ and $FE_{1,2} = [2\ |\ 1\ 3]$.

$FE_{1,1} = [1\ |\ 2\ 3]$: The primary unit of type 1 fails at time T_1 before t. The primary unit of type 2 and the cold unit of type 3 powered at time T_1 survive mission time t.

$$\Pr\{FE_{1,1}\} = \int_0^t f_{T_1}(t_1)\left(1 - \int_{t_1}^t f_{T_3}(t_3 - t_1)dt_3\right)dt_1(1 - \int_0^t f_{T_2}(t_2)dt_2) = 0.0905$$

Similarly, $\Pr\{FE_{1,2}\}$ can be obtained as:

$$\Pr\{FE_{1,2}\} = \int_0^t f_{T_2}(t_1)(1 - \int_{t_1}^t f_{T_3}(t_3 - t_1)dt_3)dt_1(1 - \int_0^t f_{T_1}(t_2)dt_2) = 0.1989$$

(c) Event FE_2: Two units fail before t. There are four possible failure sequences ($f_2 = 4$): $FE_{2,1} = [1\ 2\ |\ 3]$, $FE_{2,2} = [2\ 1\ |\ 3]$, $FE_{2,3} = [1\ 3\ |\ 2]$, $FE_{2,4} = [2\ 3\ |\ 1]$.

$FE_{2,1} = [1\ 2\ |\ 3]$: The primary unit of type 1 fails at time T_1, then the primary unit of type 2 fails at time (T_2) before t and the cold unit of type 3 powered at time T_1 survives mission time t.

$$\Pr\{FE_{2,1}\} = \int_0^t \int_{t_1}^t f_{T_1}(t_1)f_{T_2}(t_2)(1 - \int_{t_1}^t f_{T_3}(t_3 - t_1)dt_3)dt_2dt_1 = 0.0193$$

Similarly, $\Pr\{FE_{2,2}\}$, $\Pr\{FE_{2,3}\}$, and $\Pr\{FE_{2,4}\}$ are obtained as:

$$\Pr\{FE_{2,2}\} = \int_0^t \int_{t_1}^t f_{T_2}(t_1)f_{T_1}(t_2)(1 - \int_{t_1}^t f_{T_3}(t_3 - t_1)dt_3)dt_2dt_1 = 0.0206$$

$$\Pr\{FE_{2,3}\} = \int_0^t \int_{t_1}^t f_{T_1}(t_1)f_{T_3}(t_2 - t_1)dt_2dt_1(1 - \int_0^t f_{T_2}(t_3)dt_3) = 0.0310$$

$$\Pr\{FE_{2,4}\} = \int_0^t \int_{t_1}^t f_{T_2}(t_1) f_{T_3}(t_2 - t_1) dt_2 dt_1 (1 - \int_0^t f_{T_1}(t_3) dt_3) = 0.07095$$

(d) Event FE_3: Three units fail before t. There are four possible failure sequences ($f_3 = 4$): $FE_{3,1} = [1\ 2\ 3]$, $FE_{3,2} = [2\ 1\ 3]$, $FE_{3,3} = [1\ 3\ 2]$, $FE_{3,4} = [2\ 3\ 1]$

$FE_{3,1} = [1\ 2\ 3]$: The primary unit of type 1 fails at time T_1, then the primary unit of type 2 fails at time (T_2) and then the powered cold unit of type 3 fails at time (T_1+T_3) before t.

$$\Pr\{FE_{3,1}\} = \int_0^t \int_{t_1}^t \int_{t_2}^t f_{T_1}(t_1) f_{T_2}(t_2) f_{T_3}(t_3 - t_1) dt_3 dt_2 dt_1 = 0.0047$$

Similarly, $\Pr\{FE_{3,2}\}$, $\Pr\{FE_{3,3}\}$, and $\Pr\{FE_{3,4}\}$ are evaluated as:

$$\Pr\{FE_{3,2}\} = \int_0^t \int_{t_1}^t \int_{t_2}^t f_{T_2}(t_1) f_{T_1}(t_2) f_{T_3}(t_3 - t_1) dt_3 dt_2 dt_1 = 0.0049$$

$$\Pr\{FE_{3,3}\} = \int_0^t \int_{t_1}^t \int_{t_2}^t f_{T_1}(t_1) f_{T_3}(t_2 - t_1) f_{T_2}(t_3) dt_3 dt_2 dt_1 = 0.0049$$

$$\Pr\{FE_{3,4}\} = \int_0^t \int_{t_1}^t \int_{t_2}^t f_{T_2}(t_1) f_{T_3}(t_2 - t_1) f_{T_1}(t_3) dt_3 dt_2 dt_1 = 0.0054$$

For this system design, system unreliability $Q(t) = \sum_{i=2}^3 \sum_{p=1}^{f_i} \Pr\{FE_{i,p}\}$ $= 0.162$. $R(t) = 1 - Q(t) = 0.838$.

Step 2: The energy consumption equation for each $FE_{i,p}$:

$$E_{0,1}(t) = r_1 t + r_2 t + e_{01} + e_{02}$$
$$E_{1,1}(t) = r_1 T_1 + r_2 t + r_3(t - T_1) + e_{01} + e_{02} + e_{03}$$
$$E_{1,2}(t) = r_2 T_1 + r_1 t + r_3(t - T_1) + e_{01} + e_{02} + e_{03}$$
$$E_{2,1}(t) = r_1 T_1 + r_2 T_2 + r_3(t - T_1) + e_{01} + e_{02} + e_{03}$$
$$E_{2,2}(t) = r_2 T_1 + r_1 T_2 + r_3(t - T_1) + e_{01} + e_{02} + e_{03}$$
$$E_{2,3}(t) = r_1 T_1 + r_3 T_2 + r_2 t + e_{01} + e_{02} + e_{03}$$
$$E_{2,4}(t) = r_2 T_1 + r_3 T_2 + r_1 t + e_{01} + e_{02} + e_{03}$$
$$E_{3,1}(t) = r_1 T_1 + r_2 T_2 + r_3 T_3 + e_{01} + e_{02} + e_{03}$$
$$E_{3,2}(t) = r_2 T_1 + r_1 T_2 + r_3 T_3 + e_{01} + e_{02} + e_{03}$$
$$E_{3,3}(t) = r_1 T_1 + r_3 T_2 + r_2 T_3 + e_{01} + e_{02} + e_{03}$$
$$E_{3,4}(t) = r_2 T_1 + r_3 T_2 + r_1 T_3 + e_{01} + e_{02} + e_{03}$$

Step 3: The total expected energy consumption of the example cold-standby system with design of $V = [1\,2\,3]$ is given by

$$\overline{E(t)} = \sum_{i=0}^{3} \sum_{p=1}^{f_i} \Pr\{FE_{i,p}\} * \overline{E_{i,p}(t)}$$

Substituting the given input parameters into the above equation, we get $\overline{E(t)} = 1{,}779.8$ Wh.

Similarly, $\overline{E(t)}$ and $R(t)$ for the other two system designs [1 3 2], and [3 2 1] can be derived. The analysis results are summarized in Table 1.

Table 1: Expected energy consumption and reliability for different designs with exponential distributions

System Design V	$\overline{E(t)}$ (Wh)	$R(t)$
[1 2 3]*	1,779.8	0.838
[1 3 2]	2,328	0.829
[3 2 1]	1,903.6	0.802

*: Optimal design

Observing the analysis results in Table 1 for the three different element sequences for the 2-out-of-3:G cold-standby system with exponential time-to-failure distributions of the components choices, we can conclude that all the three sequences satisfy the system reliability constraint, i.e., $R(t) \geq 0.7$, and sequence $V = [1\,2\,3]$ has the minimum value of $\overline{E(t)}$ for mission time $t = 200$ hrs.

The proposed approach has no limitation on the component time-to-failure distribution. As an illustration, consider the case of Weibull distribution, when the *pdf* functions for all the primary/cold-standby units are substituted with $\frac{k}{\lambda}\left(\frac{t}{\lambda}\right)^{k-1} e^{-\left(\frac{t}{\lambda}\right)^{k}}$ with shape parameter k and scale parameter λ. The mean of the random variable following the Weibull distribution is $\lambda\Gamma\left(1 + \frac{1}{k}\right)$, where $\Gamma(z)$ is a gamma function given $\Gamma(z) = \int_{0}^{\infty} x^{z-1} e^{-x} dx$.

Assume the given set of three components have non-identical power rates of $r_1 = 6$ W, $r_2 = 4$ W, $r_3 = 2$ W, start-up energy consumption of $e_{01} = 7$ Wh, $e_{02} = 5$ Wh, $e_{03} = 3$ Wh, shape parameters of $k_1 = 1$, $k_2 = 2$, $k_3 = 2.5$, and scale parameters of $\lambda_1 = 100$, $\lambda_2 = 150$, $\lambda_3 = 200$. The $\overline{E(t)}$ for all three system designs [1 2 3], [1 3 2], and [3 2 1] can be derived using the method described in Section 3.1 for mission time $t = 100$ hrs. The obtained results are summarized in Table 2.

Observing the analysis results in Table 2 for the three different sequences for Weibull time-to-failure distributions of the components choices, we can conclude that the sequence V = [3 2 1] has the minimum value of $\overline{E(t)}$ for mission time t = 100 hrs, while satisfying the constraint on system reliability $R(t) \geq 0.7$.

Table 2: Expected energy consumption and reliability for different sequences with Weibull distributions

System Design V	$\overline{E(t)}$ (Wh)	R(t)
[1 2 3]	1,063	0.746
[1 3 2]	810.4	0.809
[3 2 1]*	672.2	0.831

*: Optimal design

In this 2-out-of-3:G example system, there is only one cold-standby unit in the system design, therefore, the effect of the cold-standby unit's initiation sequence cannot be illustrated. Next, the effect of the initiation sequence is explained through a 2-out-of-4:G example system.

Example 2: 2-out-of-4:G cold-standby system (n = 4, and k = 2)

Assume the given set of four components contains components of type 1, 2, 3 and 4, respectively. They have non-identical power rates of r_1 = 5 W, r_2 = 4 W, r_3 = 6 W, r_4 = 4.5 W, failure rates of λ_1= 0.001/hr, λ_2 = 0.002/hr, λ_3 = 0.003/hr, λ_4 = 0.004/hr and start-up energy consumption of e_{01} = 6 Wh, e_{02} = 5 Wh, e_{03} = 7 Wh, e_{04} = 5 Wh. With the given set of components, there are in total twelve possible 2-out-of-4 cold-standby system designs: [1 2 3 4], [1 3 2 4], [2 3 1 4], [1 2 4 3], [1 3 4 2], [1 4 2 3], [1 4 3 2], [2 3 4 1], [2 4 1 3], [2 4 3 1], [3 4 1 2], and [3 4 2 1]. The system fails when $n-k+1(=3)$ components fail. As an illustration, in the following, we evaluate $\overline{E(t)}$ for each of the system designs with detailed illustration for V = [1 2 3 4]. Assume mission time t = 100 hrs and R^*= 0.9.

Solution for the cold-standby system V = [1 2 3 4]: Meaning the system has two primary units of type 1 and type 2 and two cold-standby units of type 3 and type 4. Initially the primary units of type 1 and type 2 are online. When any one of the primary unit (say type 2) has failed, it is replaced with the cold unit of type 3 in the sequence. When any one of the online units type 1 or type 3 (say type 3) has failed, it is replaced with the cold unit of type 4 in the sequence. The system fails when either primary unit of type 1 or powered cold unit of type 4 has failed.

Step 1: Construct an *FE* space. There are five possible failure events FE_i, where i = 0, 1, 2, 3, 4.

(a) Event FE_0: No unit fails before t. There is only one possible combination $(f_0 = 1)$, i.e., all $k = 2$ primary components survive the mission time. In particular,

$FE_{0,1} = [1\ 2]$: The primary units of type 1 and type 2 survive mission time t.

$$\Pr\{FE_{0,1}\} = \left(1 - \int_0^t f_{T_1}(t_1)\,dt_1\right)\left(1 - \int_0^t f_{T_2}(t_2)\,dt_2\right)$$

(b) Event FE_1: One unit fails before t. There are two possible failure sequences $(f_1 = 2)$: $FE_{1,1} = [1\ |\ 2\ 3]$ and $FE_{1,2} = [2\ |\ 1\ 3]$. For example,

$FE_{1,1} = [1\ |\ 2\ 3]$: The primary unit of type 1 fails at time T_1 before t. The primary unit of type 2 and the cold unit of type 3 powered at time T_1 survive mission time t.

$$\Pr\{FE_{1,1}\} = \int_0^t f_{T_1}(t_1)\left(1 - \int_{t_1}^t f_{T_3}(t_3 - t_1)\,dt_3\right)dt_1(1 - \int_0^t f_{T_2}(t_2)\,dt_2)$$

$\Pr\{FE_{1,2}\}$ can be similarly evaluated.

(c) Event FE_2: Two units fail before t. There are four possible failure sequences $(f_2 = 4)$: $FE_{2,1} = [1\ 2\ |\ 3\ 4]$, $FE_{2,2} = [2\ 1\ |\ 3\ 4]$, $FE_{2,3} = [1\ 3\ |\ 2\ 4]$, $FE_{2,4} = [2\ 3\ |\ 1\ 4]$. For example,

$FE_{2,1} = [1\ 2\ |\ 3\ 4]$: The primary unit of type 1 fails at time T_1, then the primary unit of type 2 fails at time T_2 before t and the cold units of type 3 powered at time T_1 and type 4 powered at time T_2 survive mission time t.

$$\Pr\{FE_{2,1}\} = \int_0^t \int_{t_1}^t f_{T_1}(t_1)f_{T_2}(t_2)\left(1 - \int_{t_1}^t f_{T_3}(t_3 - t_1)\,dt_3\right)\left(1 - \int_{t_2}^t f_{T_4}(t_4 - t_2)\,dt_4\right)dt_2dt_1$$

(d) Event FE_3: Three units fail before t. There are eight possible failure sequences $(f_3 = 8)$: $FE_{3,1} = [1\ 2\ 3\ |\ 4]$, $FE_{3,2} = [2\ 1\ 3\ |\ 4]$, $FE_{3,3} = [1\ 3\ 2\ |\ 4]$, $FE_{3,4} = [2\ 3\ 1\ |\ 4]$, $FE_{3,5} = [1\ 2\ 4\ |\ 3]$, $FE_{3,6} = [2\ 1\ 4\ |\ 3]$, $FE_{3,7} = [1\ 3\ 4\ |\ 2]$, $FE_{3,8} = [2\ 3\ 4\ |\ 1]$. For example,

$FE_{3,1} = [1\ 2\ 3\ |\ 4]$: The primary unit of type 1 fails at time T_1, then the primary unit of type 2 fails at time T_2, the powered cold unit of type 3 fails at time $(T_1 + T_3)$ before t and the cold unit type 4 powered at time T_2 survives mission time t.

$$\Pr\{FE_{3,1}\} = \int_0^t \int_{t_1}^t \int_{t_2}^t f_{T_1}(t_1)f_{T_2}(t_2)f_{T_3}(t_3 - t_1)\left(1 - \int_{t_2}^t f_{T_4}(t_4 - t_2)\,dt_4\right)dt_3dt_2dt_1$$

(e) Event FE_4: All units fail before t. There are eight possible failure sequences ($f_4 = 8$): $FE_{4,1} = [1\ 2\ 3\ 4]$, $FE_{4,2} = [2\ 1\ 3\ 4]$, $FE_{4,3} = [1\ 3\ 2\ 4]$, $FE_{4,4} = [2\ 3\ 1\ 4]$, $FE_{4,5} = [1\ 2\ 4\ 3]$, $FE_{4,6} = [2\ 1\ 4\ 3]$, $FE_{4,7} = [1\ 3\ 4\ 2]$, $FE_{4,8} = [2\ 3\ 4\ 1]$. For example,

$FE_{4,1} = [1\ 2\ 3\ 4]$: The primary unit of type 1 fails at time T_1, then the primary unit of type 2 fails at time T_2, the powered cold unit of type 3 fails at time (T_1+T_3) and then the powered cold unit type 4 fails at time (T_2+T_4) before t.

$$\Pr\{FE_{4,1}\} = \int_0^t \int_{t_1}^t \int_{t_2}^t \int_{t_3}^t f_{T_1}(t_1) f_{T_2}(t_2) f_{T_3}(t_3 - t_1) f_{T_4}(t_4 - t_2) dt_4 dt_3 dt_2 dt_1$$

Step 2: Develop energy consumption equations. For example,

$E_{0,1}(t) = r_1 t + r_2 t + e_{01} + e_{02}$

$E_{1,1}(t) = r_1 T_1 + r_2 t + r_3(t-T_1) + e_{01} + e_{02} + e_{03}$

$E_{2,4}(t) = r_1 T_1 + r_2 T_2 + r_3(t-T_1) + r_4(t-T_2) + e_{01} + e_{02} + e_{03} + e_{04}$

$E_{3,1}(t) = r_1 T_1 + r_2 T_2 + r_3 T_3 + r_4(t-T_2) + e_{01} + e_{02} + e_{03} + e_{04}$

$E_{4,1}(t) = r_1 T_1 + r_2 T_2 + r_3 T_3 + r_4 T_4 + e_{01} + e_{02} + e_{03} + e_{04}$

Step 3: The total expected energy consumption of the example cold-standby system $V = [1\ 2\ 3\ 4]$ is given by

$$\overline{E(t)} = \sum_{i=0}^{4} \sum_{p=1}^{f_i} \Pr\{FE_{i,p}\} * \overline{E_{i,p}(t)}$$

Substituting the given input parameters into the above equation, we get $\overline{E(t)} = 792.29\ Wh$ and $R(t) = 0.9909$.

Similarly, $\overline{E(t)}$ and $R(t)$ for the other eleven system designs can be derived. The analysis results are summarized in Table 3. Observing the analysis results in Table 3 for the twelve different element sequences for the 2-out-of-4:G cold-standby system, we can conclude that sequence $V = [2\ 4\ 3\ 1]$ (two primary units of type 2 and type 4 and two cold-standby units of type 3 and type 1 with initiation sequence of 3 and then 1) offers the minimum expected total system energy consumption while satisfying constraint of $R(t) \geq 0.9$ for $t = 100$ hrs. We also verify that both reliability and energy consumption of a k-out-of-n:G cold-standby system depend on the selection of versions of system components for primary and cold-standby components and on the initiation sequence of the cold-standby components.

Table 3: Expected energy consumption and reliability for different designs with exponential distributions

System Design V	$\overline{E(t)}$ (Wh)	R(t)
[1 2 3 4]	792.29	0.9909
[1 2 4 3]	902.84	0.9905
[1 3 4 2]	1,316.63	0.9892
[1 3 2 4]	1,342.85	0.9905
[1 4 2 3]	1,035.55	0.9899
[1 4 3 2]	952.82	0.9891
[2 3 1 4]	1,067.17	0.9891
[2 3 4 1]	1,094.4	0.9855
[2 4 1 3]	780.55	0.9886
[2 4 3 1]*	753.59	0.9853
[3 4 1 2]	1,193.32	0.9874
[3 4 2 1]	1,214.51	0.9850

*: Optimal design

4. Case studies

The proposed methodology is further implemented on an example data set (Table 4) for two different cases. The brute force approach is implemented, where all the possible design solutions are evaluated and those offering the best value of $\overline{E(t)}$ and satisfying the system reliability constraint are chosen as the final optimal system designs.

Table 4: Example data set

Component Type i	λ_i (/hr)	e_{0i}	r_i
1	0.001	6	5
2	0.002	5	4
3	0.003	7	6
4	0.004	5	4.5
5	0.005	4	3
6	0.006	4	3.5

Table 5 presents the optimal design results for the example data set of Table 4 for the following two cases.

Case I: $n = 5$ with given component choices of types 1, 2, 3, 4 and 5;
Case II: $n = 6$ with given component choices of types 1, 2, 3, 4, 5 and 6.

Table 5: Optimal design results

Case n		k	R^*	t	Optimal solution	$\overline{E(t)}$ (Wh)	$R(t)$
I	5	2	0.9	100	[2 5 3 1 4]	524.42	0.9977
II	6	3	0.9	100	[2 5 6 3 1 4]	904.4	0.9821

The running times of the Matlab code realization of the proposed method on 2.2 GHz i-7 core processor for Case I is about 6 hours and for Case II is about 35 hours. Note that the example 2 in Section 3.2 corresponds to the case where $n = 4$ with component choices of types 1, 2, 3, and 4 in Table 4. The running time for this case ($n = 4$) is about 20 minutes.

5. Conclusions and future work

In this chapter, the optimal standby element sequencing problem (SESP) for heterogeneous k-out-of-n:G cold-standby systems is considered. The objective is to distribute n non-identical component choices between k primary online components and $(n-k)$ cold-standby components and to determine the initiation sequence of all the cold-standby components for minimizing the expected system energy consumption while satisfying a constraint on the entire system reliability. An analytical method is proposed for the energy consumption and reliability analysis of the k-out-of-n:G cold-standby system. The brute force approach is applied in order to obtain the optimal design configuration.

Our future work includes solving the SESP for 1-out-of-n:G and k-out-of-n:G heterogeneous standby systems with mixed hot-cold redundancy types considering recovery time and operational cost based on energy consumption and reliability of the system.

REFERENCES

Boddu, P. and L. Xing. 2011. Optimal design of heterogeneous series-parallel systems with common cause failures. International Journal of Performability Engineering, special Issue on Performance and Dependability Modeling of Dynamic Systems 7(5): 455–466.

Boddu, P. and L. Xing. 2012. Redundancy allocation for k-out-of-n:G systems with mixed spare types. Proceedings of the 58th Annual Reliability & Maintainability Symposium, Reno, NV, USA.

Boddu, P., L. Xing and G. Levitin. 2014. Energy consumption modeling and optimization in heterogeneous cold-standby systems. International Journal of Systems Science: Operations & Logistics 1(3): 142–152.

Boddu, P., L. Xing and O. Tannous. 2011. Optimal design of heterogeneous series-parallel systems with mixed redundancy types. Proceedings of the 7th International Conference on Mathematical Methods in Reliability 99–105.

Cao, D., S. Kan and Y. Sun. 2012. Design of reliable system based on dynamic Bayesian networks and genetic algorithm. Proceedings of the 58th Annual Reliability & Maintainability Symposium, Reno, NV, USA.

Chen, T.-C. and P.-S. You. 2005. Immune algorithms-based approach for redundant reliability problems with multiple component choices. Computers in Industry 56(2): 195–205.

Chern, M.S. 1992. On the computational complexity of reliability redundancy allocation in a series system. Operations Research Letters 11: 309–315.

Coit, D.W. 2001. Cold-standby redundancy optimization for non-repairable systems. IIE Transactions 33: 471–478.

Coit, D.W. 2003. Maximization of system reliability with a choice of redundancy strategies. IIE Transactions 35(6): 535–543.

Coit, D.W. and J. Liu. 2000. System reliability optimization with k-out-of-n subsystems. International Journal of Reliability, Quality and Safety Engineering 7(2): 129–142.

Coit, D.W. and A.E. Smith. 1996. Reliability optimization of series-parallel systems using a genetic algorithm. IEEE Transactions on Reliability 45(2): 254–260.

Dai, Y., G. Levitin and L. Xing. 2017. Optimal periodic inspections and activation sequencing policy in standby systems with condition based mode transfer. IEEE Transactions on Reliability 66(1): 189–201.

Fyffe, D.E., W.W. Hines and N.K.Lee. 1968. System reliability allocation and a computation algorithm. IEEE Transactions on Reliability 17: 64–69.

Gen, M. and Y. Yun. 2006. Soft computing approach for reliability optimization: Sstate-of-the-art survey. Reliability Engineering and System Safety 91(9): 1008–1026.

Kuo, W., V.R. Prasad, F.A. Tillman and C. Hwang. 2001. Optimal Reliability Design Fundamental and Application. Cambridge University Press, London.

Kuo, W. and R. Wan. 2007. Recent advances in optimal reliability allocation. IEEE Transactions on Systems, Man and Cybernetics, Part A: Systems and Humans 37(2): 43–156.

Levitin, G., L. Xing and Y. Dai. 2013a. Cold-standby sequencing optimization considering mission cost. Reliability Engineering and System Safety 118: 28–34.

Levitin, G., L. Xing and Y. Dai. 2013b. Sequencing optimization in k-out-of-n cold-standby systems considering mission cost. International Journal of General Systems 42(8): 870–882.

Levitin, G., L. Xing and Y. Dai. 2013c. Optimal sequencing of warm standby elements. Computers & Industrial Engineering 65(4): 570–576.

Levitin, G., L. Xing and Y. Dai. 2014a. Cold vs. hot standby mission operation cost minimization for 1-out-of-N Systems. European Journal of Operational Research 234(1): 155–162.

Levitin, G., L. Xing and Y. Dai. 2014b. Mission cost and reliability of 1-out-of-n warm standby systems with imperfect switching mechanisms. IEEE Transactions on Systems, Man, and Cybernetics: Systems 44(9): 1262–1271.

Levitin, G., L. Xing and Y. Dai. 2014c. Optimization of predetermined standby mode transfers in 1-out-of-N:G systems. Computers and Industrial Engineering 72: 106–113.

Levitin, G., L. Xing and Y. Dai. 2015a. Effect of failure propagation on cold vs. hot standby tradeoff in heterogeneous 1-out-of-N:G systems. IEEE Transactions on Reliability 64(1): 410–419.

Levitin, G., L. Xing and Y. Dai. 2015b. Reliability and mission cost of 1-out-of-N:G systems with state-dependent standby mode transfers. IEEE Transactions on Reliability 64(1): 454–462.

Levitin, G., L. Xing and Y. Dai. 2015c. Optimal design of hybrid redundant systems with delayed failure-driven standby mode transfer. IEEE Transactions on Systems, Man, and Cybernetics: Systems 45(10): 1336–1344.

Levitin, G., L. Xing and Y. Dai. 2015d. Heterogeneous 1-out-of-N warm standby systems with dynamic uneven backups. IEEE Transactions on Reliability 64(4): 1325–1339.

Levitin, G., L. Xing and Y. Dai. 2015e. Optimal backup distribution in 1-out-of-N cold standby systems. IEEE Transactions on Systems, Man, and Cybernetics: Systems 45(4): 636–646.

Levitin, G., L. Xing and Y. Dai. 2016a. Cold-standby systems with imperfect backup. IEEE Transactions on Reliability 65(4): 1798–1809.

Levitin, G., L. Xing and Y. Dai. 2018. Heterogeneous 1-out-of-N warm standby systems with online checkpointing. Reliability Engineering & System Safety 169: 127–136.

Levitin, G., L. Xing, Y. Dai and V.M. Vokkarane. 2017. Dynamic checkpointing policy in heterogeneous real-time standby systems. IEEE Transactions on Computers 66(8): 1449–1456.

Levitin, G., L. Xing, B.W. Johnson and Y. Dai. 2015. Mission reliability, cost and time for cold standby computing systems with periodic backup. IEEE Transactions on Computers 64(4): 1043–1057.

Levitin, G., L. Xing, Q. Zhai and Y. Dai. 2016b. Optimization of full vs. incremental periodic backup policy. IEEE Transactions on Dependable and Secure Computing 13(6): 644–656.

Liang, Y.C. and A.E. Smith. 2004. An ant colony optimization algorithm for the redundancy allocation problem (RAP). IEEE Transactions on Reliability 53(3): 417–423.

Misra, K.B. 1972. Reliability optimization of series-parallel system. IEEE Transactions on Reliability 21: 230–238.

Misra, K.B. and U. Sharma. 1971. An efficient algorithm to solve integer programming problems arising in system-reliability design. IEEE Transactions on Reliability 40(1): 81–91.

Onishi, J., S. Kimura, R.J.W. James and Y. Nakagawa. 2007. Solving the redundancy allocation problem with a mix of components using the improved surrogate constraint method. IEEE Transactions on Reliability 56(1): 94–101.

Tavakkoli-Moghaddam, R., J. Safari and F. Sassani. 2008. Reliability optimization of series-parallel systems with a choice of redundancy strategies using a genetic algorithm. Reliability Engineering and System Safety 93(4): 550–556.

Xing, L. and S.V. Amari. 2008. Fault tree analysis. Chapter 38. *In*: Krishna B. Misra (ed.). Handbook of Performability Engineering. Springer-Verlag. ISBN: 978-1-84800-130-5.

Impact of Correlated Failure on the Maintenance of Multi-state Consecutive 2-out-of-n: Failed Systems

Bentolhoda Jafary* and Lance Fiondella

University of Massachusetts-Dartmouth, 285 Old Westport Rd, North Dartmouth, MA 02747, USA

Notations

n	Number of components
k_j	Number of distinct non-zero performance levels of component j
g_{jh}	Performance of component j in state h
g_j	Vector of performance levels of component j
G_j	Random variable of performance level of component j
p_{jh}	Probability component j exhibits state h
p_j	Vector of component j states
φ	Mapping of component performances to system performance, $L^n \to M$
W	Random variable of system performance level
w_i	System performance level in state $i \in \{0, 1, ..., K\}$
q_i	Probability system exhibits performance level w_i
$R^x_{w_{min}}$	Probability system x performance meets or exceeds minimum threshold w_{min}
u^i	u-function of component i
$\otimes_{\varphi X}$	Composition operator of u-functions for system $x \in X$

---

*Corresponding author: bjafary@umassd.edu

ρ Component correlation

C_{ER} Cost of emergency repair under age replacement to minimize cost

C_{PM} Preventive maintenance cost under age replacement to minimize cost

$\eta_{age}(T)$ Average cost per unit time under age replacement to minimize cost

t_{ER} Time to perform emergency repair under age replacement to maximize availability

t_{PR} Time to perform preventive repair under age replacement to maximize availability

$\eta_F(T)$ Average availability per unit time under age replacement to maximize availability

1. Introduction

Most reliability models assume that the failures of the components of a system are statistically independent (Trivedi 2008). This assumption simplifies the calculations. In practice (Dhillon and Anude 1994), however, a system can experience correlated failure. This is especially true of consecutive k-out-of-n failed and multi-state systems. The MSS paradigm can model systems composed of components that exhibit three or more levels of performance. MSS can experience various kinds of dependency and correlated failure such as simultaneous or cascading failure, which can negatively effect system reliability and mean time to failure (MTTF).

Applications and models of multi-state consecutive k-out-of-n:F systems include the work of Hamim and Porat (Haim and Porat 1991) who proposed a Bayesian reliability model to describe a multi-state consecutive k-out-of-n:F system with independent and identical components and derived mathematical expressions for system unreliability. Huang et al. (2000) presented a multi-state consecutive k-out-of-n:G system model, extending an algorithm to evaluate system reliability from binary to MS systems. Later, Huang et al. (Huang et al. 2003) proposed definitions of multi-state consecutive k-out-of-n:F and G systems, identifying the dual relationship between the two. The authors provided an algorithm to evaluate the system state distribution of decreasing multi-state consecutive k-out-of-n:F systems. Belaloui and Ksir (2007) modelled multi-state consecutive k-out-of-n:G systems, providing a recursive-formula and expressions for the reliability of decreasing, increasing, and constant multi-state consecutive k-out-of-n:G systems. Habib et al. (2007) proposed a model to generalize the linear consecutive k-out-of-n-from-n:G system from binary to multi-state. The authors also evaluated the reliability of the special case of multi-state consecutive k-out-of-r-from-n:G systems, providing an algorithm based on the total probability theorem. Levitin

(Levitin 2013) introduced a multistate vector k-out-of-n:F system model and proposed an algorithm to evaluate the system's state probabilities. Mo et al. (2017) proposed a multi-valued decision diagram (MDD) approach to model and evaluate performability of a heterogeneous multi-state linear consecutive k-out-of-n:F system with arbitrary life distribution.

Nourelfath and Ait-Kadi (2007) proposed a method based on a combination of the universal generating function (UGF) and the Markov model to evaluate the availability of repairable modules under maintenance restrictions considering dependencies that result from shared maintenance teams. Liu and Huang (2010) introduced an optimal replacement policy for multi-state systems possessing multi-state elements whose aging process is characterized by a non-homogeneous continuous time Markov model in order to identify the optimal number of failures after which a system should be replaced. Gurler and Kaya (2002) considered a multi-component multi-state system model where each component is classified as good, doubtful, preventive maintenance due, and down, in which the maintenance policy is based on the number of the components in the "doubtful" states.

This chapter presents a method to characterize the impact of correlated component failures on the optimal preventive maintenance interval of a universal generating function-based (Levitin 2004) multi-state consecutive 2-out-of-n:F system consisting of identical but correlated elements. This method is applied to two maintenance policies: i) age replacement to minimize cost and ii) age replacement to maximize availability. We study the impact of correlated failures on the performance and reliability of a discrete multi-state consecutive 2-out-of-n:F system. We then generalize the discrete model to the time varying case and assess the impact of correlation in the context of two maintenance models.

The remainder of this chapter is organized as follows: Section II reviews the consecutive k-out-of-n failed system and UGF-based modeling approach and develops discrete and continuous models for MSS composed of correlated identical components. Section III reviews the two maintenance models considered. Section IV illustrates how the approach can be used to assess the impact of correlation on these discrete and continuous models, including the optimality of maintenance. Section V provides conclusions and offers directions for future research.

2. Modelling

This section provides an overview of the modelling approach. Section 2.1 introduces the consecutive k-out-of-n failed system. Sections 2.2 and 2.3 review the UGF approach to MSS assessment (Levitin 2004) for the special case when component failures are uncorrelated, respectively discussing modelling and the reliability block diagram method to obtain an expression of MSS performance from the performance distributions of the multi-state components. Section 2.4 introduces correlated identical components into the elements of a discrete multi-state component, while Section 2.5 generalizes to the continuous case and reviews related measures of interest for the continuous case.

2.1 Consecutive *k*-out-of-*n* failed system with independent component failures

The linear consecutive *k*-out-of-*n* failed system (Chang et al. 2004), commonly abbreviated to consecutive *k*-out-of-*n*:F, is a system with *n* components arranged in series. As the name suggests, such a system fails if and only if *k* consecutive components fail.

Figure 1 shows the representation of a multi-state consecutive *k*-out-of-*n*:F system in terms of its cut sets.

There are $n - k + 1$ such cut sets. Each stage consists of two multi-state components in parallel. Moreover, each component can exhibit performance level zero, one, or two. Thus, each multi-state stage can exhibit performance level zero through four. System failure occurs when the performance of *k* consecutive multi-state stages falls below a desired performance level.

2.2 Multi-state systems

In general, a system consists of *n* components with the *j*th component possessing $k_j + 1$ states. The states indicate the possible performance levels and are represented by the set $g_j = \{g_{j0}, g_{j1}, ..., g_{jk_j}\}$. Here g_{jh} denotes the performance of component *j* in state $h \in \{0, 1, ..., k_j\}$. The performance realized by component *j* is represented by the random variable, which may take values from the set g_j. The probability that the *j*th component

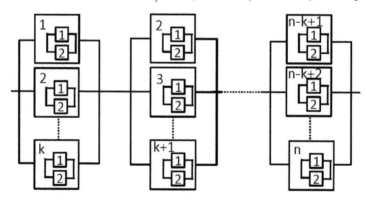

Fig. 1: Multi-state consecutive *k*-out-of-*n*:F cut sets

assumes one of the $k_j + 1$ possible states is determined by the discrete probability vector

$$p_j = \{p_{j0}, p_{j1}, ..., p_{jkj}\} \tag{1}$$

with $\quad\quad\quad p_{jh} = \text{Pr}\{G_j = g_{jh}\}$, and $\sum_{h=0}^{k_j} p_{jh} = 1.0.$

The space of possible combinations of component performances is $L^n = \{g_{10}, ..., g_{1k_1}\} \times \{g_{20}, ..., g_{2k_2}\} \times ... \times \{g_{n0}, ..., g_{nk_n}\}$. Let the entire system consist of $|L^n| = K + 1$ states with w_i representing the system performance in state $i \in 0, ..., K$, and W the random variable for system performance which assumes values from the set $M = \{w_0, ..., w_K\}$. The transform $\varphi(G_1, ..., G_n)$: $L^n \to M$ maps the space of component performances to the space of system performances and is referred to as the system structure function. From the component performance distributions (given by Equation (1) and the system structure function, it is possible to obtain the system performance distribution

$$q_i = \text{Pr}\{W = w_i\}, \quad 0 \le i \le K \tag{2}$$

where q_i is the probability that the system exhibits performance level w_i.

$$R_{w_{min}} = \text{Pr}\{W > w_{min}\} = \sum_{i=w_{min}}^{k} \text{Pr}\{W = w_i\} \tag{3}$$

where $1 \le w_{min} \le K$.

2.3 Reliability block diagram method for multi-state consecutive *k*-out-of-*n*:F system

The structure function of a MS consecutive *k*-out-of-*n*:F system consisting of series and parallel sub-systems may be determined by iteratively composing the structure functions of the independent components. Pairs of components are composed until the overall system structure function is obtained. The algorithm to produce this system performance distribution is as follows:

1. Represent the probability mass function of the random performance for each system component *j* by a *u*-function as:

$$u_{j(z)} = \sum_{h_j=0}^{k_j} p_j h_j z^{g_j h_j} \tag{4}$$

2. Identify a pair of components in series or parallel.
3. Apply the binary composition operator to the μ-functions of the identified components *i* and *j*.

$$u_{\{i,j\}}(z) = u_i(z) \otimes_{\Phi x} u_j(z) = \sum_{h_j}^{k_j} \sum_{h_j}^{k_j} p_{ihj} p_{jhj} z^{\Phi_x(g_{ih_j}, g_{jh_j})} \tag{5}$$

4. Replace components i and j with a single component produced by Equation (5).
5. If the MSS still contains more than one component, return to step 2.

The choice of ϕ in step 3 depends on the type of system considered. When the system models capacity of a series, flow is limited by the component with the lower capacity, so that

$$\phi_s\,(G_1, G_2) = \min\,\{G_1, G_2\} \tag{6}$$

For example, if $u_1(z) = p_1 z^2 + (1-p_1)\,z^0$ and $u_2(z) = p_2 z^2 + (1-p_2)\,z^0$ then
$u_1(z) \otimes_s u_2(z) = (p_1 z^2 + (1-p_1)z^0) \otimes_s (p_2 z^2 + (1-p_2)z^0) = p_1 p_2 z^2 + (p_1\,(1-p^2)$
$+ (1-p_1)\,p_2 + (1-p_1)(1-p_2))z^0$ \hfill (7)

On the other hand, a parallel pair combines throughput, so that

$$\phi_p\,(G_1, G_2) = G_1 + G_2 \tag{8}$$

2.4 Discrete multi-state consecutive *k*-out-of-*n*:F system with correlated identical elements

As the name suggests, correlated identical elements (Fiondella and Xing 2015) are identically distributed and, therefore, possess a common component reliability, denoted here as μ. However, these elements also share a common positive correlation coefficient ρ that characterizes the correlation between the failures of each pair of elements contained in a multi-state component. The probability that multi-state component G_j exhibits performance level g_{jh} is

$$p_{jh} = \Pr\left\{G_j = g_j\right\} = \binom{k_j}{h}(\mu\beta)^h\,(1-\mu\beta)^{k_j - h},\, h \geq 1 \tag{9}$$

where $1 < h \leq k_j$ of k_j elements operate reliably and

$$\beta = 1 + \rho\left(\frac{1-\mu}{\mu}\right) \tag{10}$$

In more general probabilistic language, Equation (9) expresses the probability that $1 < h \leq k_j$ of k_j coin flips are successful, where the success parameter of the Bernoulli distribution is given by $\mu\beta$ and β explicitly contains the common correlation parameter ρ as noted in Equation (10). Thus, we characterize the performance distribution of a multi-state component with identical but correlated elements as a correlated binomial distribution with common correlation parameter ρ.

The probability of performance level zero is

$$p_{j0} = \Pr\left\{G_j = g_{j0}\right\} = 1 - \frac{1}{\beta}\sum_{h=1}^{k_j}\binom{k_j}{h}(\mu\beta)^h\,(1-\mu\beta)^{(k_j - h)} \tag{11}$$

This non-standard expression for the probability of zero successes (performance level zero in the multi-state component) occurs because summing from zero to k_j in Equation (9) produces a defective distribution, where the sum of the probabilities of the events is strictly less than one. Thus, one minus the sum of one to k_j successes preserves Kolmogorov's second axiom which states that the sum of the probabilities of the events must equal one. The correctness of Equations (9) and (11) were verified with simulation techniques employed in the derivation of these expressions (Fiondella and Xing 2015).

Therefore, the discrete reliability expression for a multi-state component with correlated identical elements is

$$R_{jh_{min}} = \sum_{i=8jh_{min}}^{k_j} \binom{k_j}{h} (\mu\beta)^i (1-\mu\beta)^{(k_j-i)} \tag{12}$$

where h_{min} denotes the minimum acceptable performance level for component j.

2.5 Continuous multi-state system with correlated identical elements

This section generalizes the discrete MSS model with correlated identical elements to the continuous case and reviews important measures for continuous systems, namely the reliability and density function.

1. *Reliability:* The continuous reliability expression $R(t)$ indicates the probability that a system survives to time t and is obtained by substituting any survival function $R(t) = 1 - F(t)$ for μ in Equation (12). Thus, the reliability expression for a multi-state component where the elements are characterized by a continuous distribution with correlated but identical elements is

$$R_{jh_{min}}(t) = \sum_{i=8jh_{min}}^{k_j} \binom{k_j}{h} ((1-F(t))\beta)^i (F(t)\beta)^{(k_j-i)} \tag{13}$$

Some possible life distributions include the exponential $F(t) = 1 - e^{-\lambda t}$ and Weibull $F(t) = 1 - e^{-\lambda t\sigma}$. For example, when the life distributions follow the exponential distribution, the expression for a continuous multi-state component composed of correlated identical elements is

$$R_{jh_{min}}(t) = \sum_{i=8jh_{min}}^{k_j} \binom{k_j}{h} (e^{-\lambda t}\beta)^i ((1-e^{-\lambda t})\beta)^{(k_j-i)} \tag{14}$$

2. *Density function:* The density function $f(t)$ is the probability density function of the time to system failure and is defined in terms of the reliability as

$$f(t) = -\frac{\partial R(t)}{\partial t} \tag{15}$$

where $R(t)$ is the system reliability.

3. Maintenance models

This section reviews two maintenance models, summarized in (Gertsbakh 2000), which we use to illustrate our method of characterizing the impact of correlated component failures on optimal maintenance policies. Section 3.1 summarizes age replacement for cost, while Section 3.2 discusses age replacement for availability criterion.

3.1 Age replacement to minimize cost

In this model, a unit possesses CDF, $F(t)$ and is replaced either when it reaches age T or upon failure prior to the replacement age. Repair upon failure (emergency repair) costs c_{ER}. Preventive replacement at age T costs c_{PM} which is significantly less than c_{ER}. Each replacement completely renews the system and requires negligible time. Let τ be the lifetime of the unit and define $Z = \min(\tau, T)$, which is the random duration prior to the first renewal. Thus, this inter-renewal period possesses mean

$$E[z] = \int_0^T (1 - F(t)) dt \tag{16}$$

The mean cost during one renewal period is $F(T)c_{ER} + (1 - F(T))c_{PM}$. Therefore, the mean cost per unit time is

$$\eta_{age} = \frac{F(T)c_{ER} + (1 - F(T))c_{PM}}{\int_0^T (1 - F(t)) dt} \tag{17}$$

3.2 Age replacement to maximize availability

Similar to the model of Section 3.1, a new unit with CDF, $F(t)$ begins operation at $t = 0$ and is replaced at age T, or failure, whichever occurs first. Preventive replacement at age T lasts for duration t_{PM}, while the time to perform emergency repair after failure is t_{ER}. Commonly, t_{ER} is significantly greater than t_{PM}.

If the unit fails in the interval $(t, t + dt)$, $t < T$, the duration of the renewal period will be $X = t + t_{ER}$, which occurs with probability $f(t)dt$. Otherwise, the duration is $X = T + t_{PM}$ if the unit reaches age T, which occurs with probability $1 - F(T)$. Thus,

$$E[X] = \int_0^T (1 - F(t)) dt + t_{ER} F(T) + t_{PM} (1 - F(T)) \tag{18}$$

On average, the unit is operational in one renewal period for time $\int_0^T \left(1 - F(t)\right) dt$. Equating reward to operational time, the mean reward per unit time is the stationary system availability

$$\eta_F(T) = \frac{\int_0^T \left(1 - F(t)\right) dt}{\int_0^T \left(1 - F(t)\right) dt + t_{ER} F(T) + t_{PM} \left(1 - F(T)\right)} \tag{19}$$

which simplifies to

$$\eta_F(T) = \left(1 + \frac{t_{ER} F(T) + t_{PM} \left(1 - F(T)\right)}{\int_0^T \left(1 - F(t)\right) dt}\right)^{-1} \tag{20}$$

Thus, maximizing availability is equivalent to minimizing cost in Equation (17).

4. Illustration

This section studies the impact of correlated failures on discrete and time-varying multi-state consecutive 2-out-of-n:F systems, denoted as MS C(2, n):F system for the sake of brevity. To illustrate trends, the examples consider multi-state consecutive 2-out-of-n for n = {3, 4, 5}. Each stage consists of two MS components, each MS component consists of two elements in parallel, and each element contributes equally to performance when operational. Thus, each component can exhibit performance level zero, one, or two and each stage can exhibit performance level zero through four. We assume the elements in the second component of stages one and two are correlated and identically distributed, while all other elements are statistically independent. Example one examines the impact of correlation on the reliability of multi-state consecutive k-out-of-n failed systems. Examples two and three explore the reliability and density function of continuous multi-state consecutive k-out-of-n failed systems. Examples four and five assess the impact of correlation on the optimal preventive maintenance interval T^* of these models and perform sensitivity studies on the optimal preventive maintenance interval and the resulting optimal cost or availability.

4.1 Reliability

This example considers the reliability of a multi-state consecutive 2-out-of-n:F system for n = {3, 4, 5}. The reliability of each element in the multi-state components is set to μ = 0.9. Moreover, the correlation between the elements of the second multi-state component in stages one and two are set to ρ = {0.0, 0.5, 1.0}, while the elements of all other components are independent. This enables us to study how correlated failures between the elements of the second multi-state component influence the entire system performance.

Figures 2a-2d show the impact of correlation on the reliability of multi-state consecutive 2-out-of-n:F systems for $n = \{3, 4, 5\}$ and component reliability $\mu = 0.9$ when the lowest acceptable performance is one through four respectively. Figures. 2a-2d indicate that the reliability of the MS $C(2, 3)$:F system is higher than the MS $C(2, 4)$:F system, which is higher than the MS $C(2, 5)$:F system. This trend occurs because increasing the number of components (n) necessitates that components be more reliable for the system to preserve the same level of reliability.

Figures 2a-2d also show that increasing correlation lowers the reliability of MS $C(2, n)$:F for the case performance level one or higher, two or higher and three or higher is acceptable. These observations agree with intuition because it should be expected that correlated failures will increase the probability that two consecutive components fail. Also, increasing correlation between the failures of the elements of the second component decreases the probability that component two exhibits performance level one and increases the probabilities of performance levels zero and two. Thus, system reliability decreases. However, for the case only performance level four is acceptable, increasing correlation increases the system reliability. This also agrees with intuition because increasing the correlation increases the probability that the consecutive components survive. Thus, increasing the probability of performance level two in component two increases the probability that the system exhibits performance level four and, thus, the reliability of the system increases.

4.2 Continuous reliability

Example IV-B and IV-C consider the continuous generalization of the MS $C(2, n)$:F system.

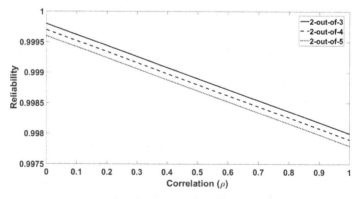

(a) Performance level one or higher acceptable

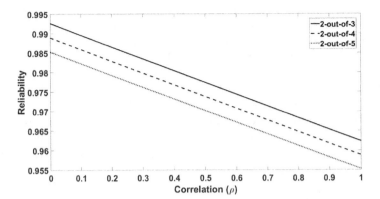

(b) Performance level two or higher acceptable

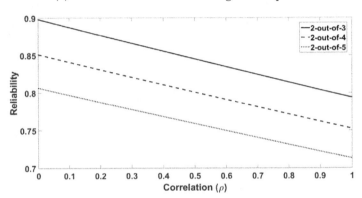

(c) Performance level three or higher acceptable

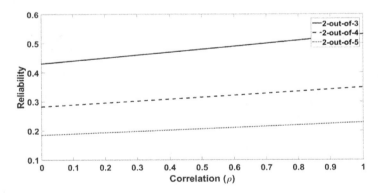

(d) Performance level four acceptable

Fig. 2: Impact of correlation on reliability of multi-state consecutive 2-out-of-n:F system

Figures 3a and 3b show the impact of correlation on the reliability of continuous MS $C(2, n)$:F systems. Figure 3a corresponds to the case where performance level one or higher is acceptable, while Fig. 3b corresponds to the case when only performance level four is acceptable. In both Figs., the upper set of three curves correspond to the continuous MS $C(2, 3)$:F system and the lower set of three curves correspond to the MS $C(2, 5)$:F system. Figures 3a and 3b indicate that the reliability of the MS $C(2, 3)$:F system is higher than the continuous MS $C(2, 4)$:F system, which is in turn higher than the continuous MS $C(2, 5)$:F system. This trend follows from the observation that, as n increases, more components must be reliable for the system to survive.

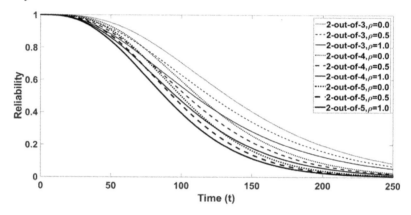

(a) Performance level one or higher acceptable

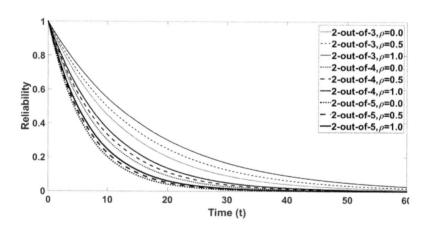

(b) Performance level four acceptable

Fig. 3: Impact of correlation on reliability of continuous multistate consecutive 2-out-of-n:F system

Figure 3a indicates that reliability of the continuous MS $C(2, n)$:F system for the case where there is no correlation between the components is highest. Increasing the correlation decreases the system reliability, as one might have anticipated, because the two elements in the second component of the system are in parallel. Increasing correlation lowers the reliability of the parallel system. Therefore, increasing correlation lowers the reliability of this second component and subsequently lowers the probability that the minimum performance threshold can be satisfied. However, Fig. 3b shows the reliability of the continuous MS $C(2, n)$:F system for the case that components are completely correlated is highest. This also agrees with intuition because performance level four is satisfied only if all of the elements perform at their highest level. in which case increasing correlation increases the probability that all of the elements works.

4.3 Density function

Figures 4a and 4b show the impact of correlation on the density function of the continuous MS $C(2, n)$:F system. Figure 4a corresponds to the case that performance level one or higher is acceptable, while Fig. 4b corresponds to the case where only performance level four is acceptable.

Figures 4a and 4b indicate that the density function of the continuous MS $C(2, 5)$:F system is higher than the continuous MS $C(2, 4)$:F and MS $C(2, 3)$:F systems. This agrees with the result shown in Fig. 3a because increasing the number of stages increases the probability of failure. Fig. ure 4a also shows that the density function of the continuous MS $C(2, n)$:F system is highest when components are completely correlated. This follows from the fact that increasing correlation increases the probability of the failure in the elements of the second components because they are in parallel. However, for the case where only performance level four is acceptable, the density function is highest when there is no correlation between the components. This also agrees with the result shown in Fig. 3b because complete correlation increases the probability that both components survive, so the density function decreases.

4.4 Age replacement to minimize cost

The goal of this model is to identify the optimal maintenance interval that minimizes the average cost per unit time. Figure 5 shows the impact of correlation on the optimal preventive maintenance interval for the MS $C(2, 3)$:F system at each of the four possible minimum performance thresholds when $c_{ER} = 10,000$, $c_{PM} = 1,000$, and $\alpha = 1.2$.

Figure 5 indicates that for $\rho = 0.0$ $T^* =$ is highest (53.3) when performance level one or higher is acceptable and that this value decreases to 12.1 when only performance level four is acceptable, which follows from the fact that

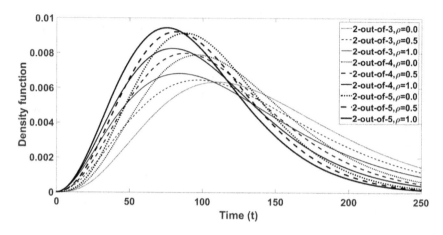

(a) Performance level one or higher acceptable

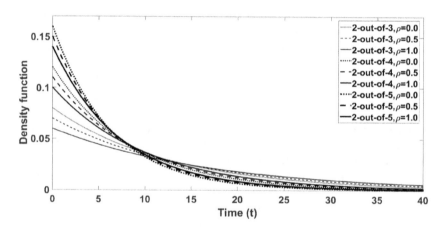

(b) Performance level four acceptable

Fig. 4: Impact of correlation on density function of continuous multi-state consecutive 2-out-of-n:F system

more of the multi-state elements must work for the minimum threshold to be satisfied. The decreasing trend is also present when $\rho = 0.5$ or $\rho = 1.0$. However, when the acceptable performance level is three or four T^* is higher for larger values of correlation.

This agrees with intuition because a higher performance level corresponds to a system where the cut set is closer to a series system and therefore benefits from correlation. Figure 6 shows the corresponding impact of correlation on the optimal cost of age replacement.

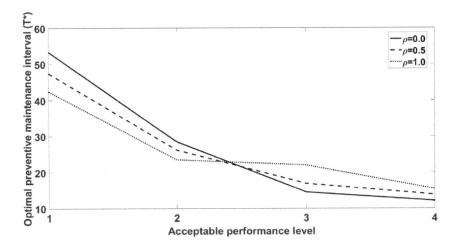

Fig. 5: Impact of correlation on optimal preventive maintenance interval (T*) under age replacement to minimize cost

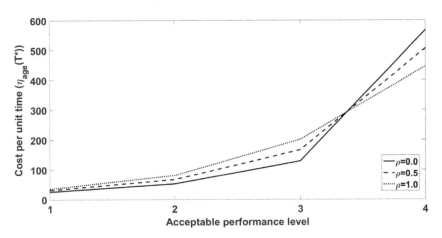

Fig. 6: Impact of correlation on age replacement to minimize cost

Figure 6 indicates that the trends in the costs are opposite to those observed in Fig. 5, as one might anticipate because larger T^* corresponds to lower cost. Table 1 summarizes the values of the optimal maintenance interval and cost for MS $C(2,n)$:F systems, $n \in \{3,4,5\}$ under age replacement to minimize cost.

Table 1 indicates that the trends in the maintenance interval and cost observed in Figs. 5 and 6 for the MS $C(2,3)$:F system are also present in the MS $C(2,4)$:F and MS $C(2,5)$:F systems.

Table 1: Impact of correlation on optimal maintenance interval and cost on MS C(2, n):F systems under age replacement to minimize cost

(a) MS consecutive 2-out-of-3

	$\rho = 0.0$		$\rho = 0.5$		$\rho = 1.0$	
	T^*	$n_{age}(T^*)$	T^*	$n_{age}(T^*)$	T^*	$n_{age}(T^*)$
P=1	53.3	25.5760	47.4	30.8855	42.4	35.3818
P=2	28.5	52.4374	26.2	66.9095	23.5	80.6891
P=3	14.5	128.9940	16.9	166.2690	22.0	201.1340
P=4	12.1	566.598	13.8	505.7080	15.4	445.8190

(b) MS consecutive 2-out-of-4

	$\rho = 0.0$		$\rho = 0.5$		$\rho = 1.0$	
	T^*	$n_{age}(T^*)$	T^*	$n_{age}(T^*)$	T^*	$n_{age}(T^*)$
P=1	47.5	28.4087	43.3	32.9665	39.6	36.9284
P=2	24.7	59.7383	22.9	73.0017	20.9	85.6131
P=3	11.9	155.0170	12.9	195.5650	14.5	236.0870
P=4	8.7	794.3610	9.4	738.0840	10.1	682.3910

(c) MS consecutive 2-out-of-5

	$\rho = 0.0$		$\rho = 0.5$		$\rho = 1.0$	
	T^*	$n_{age}(T^*)$	T^*	$n_{age}(T^*)$	T^*	$n_{age}(T^*)$
P=1	43.9	30.5547	40.6	34.6249	37.7	38.2235
P=2	22.4	65.4174	20.9	77.8049	19.3	89.5915
P=3	10.4	176.3030	10.9	218.3920	11.7	260.9860
P=4	6.8	1009.6500	7.2	956.2020	7.6	903.2460

4.5 Age replacement to maximize availability

This model seeks to maximize the steady state availability of the system per unit time. Figure 7 shows the impact of correlation on the optimal preventive maintenance interval for the MS C(2, n): F system for each performance threshold when $T_{ER} = 50$, $T_{PM} = 0.1$, and $\alpha = 1.2$.

Similar to the trends observed in Fig. 5, Fig. 7 indicates that the optimal preventive maintenance interval T^* decreases as the minimum acceptable performance level increases. Although it is not easily discernible from Fig. 7, the value of T^* is highest for $\rho = 1.0$ and lowest for $\rho = 0.0$ when the minimum acceptable performance level is four.

Figure 8 shows the corresponding impact of correlation on age replacement to maximize the availability.

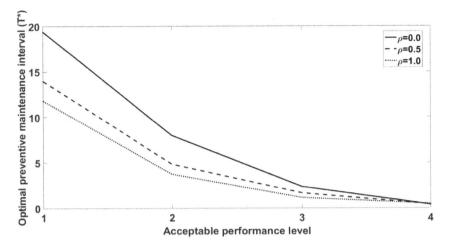

Fig. 7: Impact of correlation parameters on optimal preventive maintenance interval (T^*)

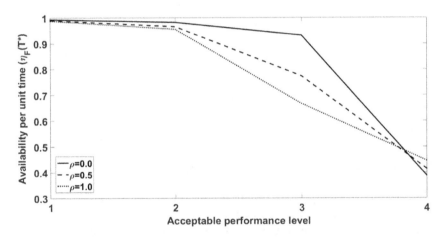

Fig. 8: Impact of correlation parameters on age replacement to maximize availability

Figure 8 indicates that the average availability per unit time $\eta_F(T^*)$ decreases as the minimum acceptable performance level increases because the maintenance interval decreases, requiring more downtime which renders the system unavailable. Similar to Table 1, Table 2 summarizes the values of the optimal time interval and average availability per unit time $\eta_F(T^*)$ for MS $C(2, n)$:F systems, $n \in \{3, 4, 5\}$ under age replacement to maximize availability.

Table 2: Impact of correlation on optimal maintenance interval and cost on MS $C(2, n)$:F systems under age replacement to maximize availability

(a) MS consecutive 2-out-of-3

	$\rho = 0.0$		$\rho = 0.5$		$\rho = 1.0$	
	T^*	$\eta_F(T^*)$	T^*	$\eta_F(T^*)$	T^*	$\eta_F(T^*)$
P=1	19.374	0.993	13.956	0.990	11.791	0.988
P=2	8.001	0.983	4.844	0.965	3.742	0.956
P=3	2.361	0.931	1.660	0.774	1.146	0.66
P=4	0.371	0.387	0.414	0.414	0.471	0.445

(b) MS consecutive 2-out-of-4

	$\rho = 0.0$		$\rho = 0.5$		$\rho = 1.0$	
	T^*	$\eta_F(T^*)$	T^*	$\eta_F(T^*)$	T^*	$\eta_F(T^*)$
P=1	17.663	0.993	13.535	0.989	11.615	0.988
P=2	7.107	0.981	4.683	0.965	3.686	0.955
P=3	1.987	0.920	1.471	0.768	1.074	0.664
P=4	0.264	0.311	0.284	0326	0.307	0.344

(c) MS consecutive 2-out-of-5

	$\rho = 0.0$		$\rho = 0.5$		$\rho = 1.0$	
	T^*	$\eta_F(T^*)$	T^*	$\eta_F(T^*)$	T^*	$\eta_F(T^*)$
P=1	16.551	0.992	13.184	0.989	11.452	0.987
P=2	6.538	0.979	4.548	0.964	3.635	0.955
P=3	1.758	0.910	1.346	0.764	1.0189	0.662
P=4	0.208	0.262	0.219	0.272	0.232	0.283

5. Conclusion and future research

This chapter developed discrete and continuous multi-state consecutive 2-out-of-n:F system models for a universal generating function-based approach where the elements comprising multi-state components are subject to identical but correlated failures. The approach was illustrated through a series of examples. The impact of correlation on reliability and optimal age replacement maintenance policies to minimize cost and maximize availability were considered. It was observed that increasing correlation can have a positive or negative influence based on different performance thresholds of the multi-state system. Thus, the approach can be used to quantify the impact of correlation on such systems.

Future work will generalize to the consecutive multi-state k-out-of-n:F systems. The impact of correlation on the hazard rate, mean time to failure and availability will also be considered.

REFERENCES

Belaloui, S. and B. Ksir. 2007. Reliability of a multi-state consecutive k-out-of-n: G system. International Journal of Reliability, Quality and Safety Engineering 14: 361–377.

Chang, G., L. Cui and F.K. Hwang. 2004. Reliabilities of Consecutive-k systems. Springer Science & Business Media, 4. USA.

Dhillon, B. and O. Anude. 1994 . Common-cause failures in engineering systems: a review. International Journal of Reliability, Quality and Safety Engineering 1: 103–129.

Fiondella, L. and L. Xing. 2015. Discrete and continuous reliability models for systems with identically distributed correlated components. Reliability Engineering and System Safety 133: 1–10.

Gertsbakh, I. 2000. Reliability theory: with applications to preventive maintenance. Springer Science and Business Media, USA.

Gürler, Ü. and A. Kaya. 2002. A maintenance policy for a system with multistate components: an approximate solution. Reliability Engineering & System Safety 76: 117–127.

Habib, A., R. Al-Seedy and T. Radwan. 2007. Reliability evaluation of multistate consecutive k-out-of-r-from-n: G system. Applied Mathematical Modelling 31: 2412–2423.

Haim, M. and Z. Porat. 1991. Bayes reliability modeling of a multistate consecutive k-out-of-n: F system, 582–586. The Proceedings of Annual Reliability and Maintainability Symposium, Piscataway. NJ, USA.

Huang, J., M.J. Zuo and Y. Wu. 2000. Generalized multi-state k-out-of-n: G systems. IEEE Transactions on Reliability 49: 105–111.

Huang, J., M.J. Zuo and Z. Fang. 2003. Multi-state consecutive-k-out-of-n systems. IIE Transactions 35: 527–534.

Levitin, G. 2004. A universal generating function approach for the analysis of multi-state systems with dependent elements. Reliability Engineering and System Safety 84: 285–292.

Levitin, G. 2013. Multi-state vector-k-out-of-n systems. IEEE Transactions on Reliability 62: 648–657.

Liu, Y. and H.Z. Huang. 2010. Optimal replacement policy for multi-state system under imperfect maintenance. IEEE Transactions on Reliability 59: 483–495.

Mo, Y., L. Xing, L. Cui and S. Si. 2017. MDD-based performability analysis of multi-state linear consecutive k-out-of-n: F systems. Reliability Engineering & System Safety 166: 124–131.

Nourelfath, M. and D. Ait-Kadi. 2007. Optimization of series-parallel multistate systems under maintenance policies. Reliability Engineering & System Safety 92: 1620–1626.

Trivedi, K.S. 2008. Probability and Statistics with Reliability, Queuing and Computer Science Applications. John Wiley & Sons, USA.

Structural and Probabilistic Examinations of *k*-out-of-*n*:G Systems and Their Application to an Optimal Construction of a Safety Monitoring System

Fumio Ohi

Emeritus Professor, Nagoya Institute of Technology, Gokiso-cho,
Syowa-ku, Nagoya, 466-8555, Japan
Visiting Researcher, Institute of Consumer Sciences and Human Life,
Kinjo Gakuin University, 1723 Omori 2-chome Moriyama-Ku
Nagoya City, 463-8521, Japan
Email: eyi06043@nitech.jp

1. Introduction

In the reliability research field, one of the most important theoretical problems is to explain ordered set theoretical and probabilistic relations among components and a system, by which useful structural and probabilistic evaluation methods have been proposed to make various large-scale systems in the modern society safe and secure.

Many basic works about binary state systems have already been performed and applied to solve practical reliability problems (Mine 1959, Birnbaum and Esary 1965, Birnbaum et al. 1961, Esary and Proschan 1963, Bodin 1970). Barlow and Proschan (1975) have summarized these works in their book.

A system and its components, however, could take intermediate performance states between functioning and failure states, and some states may not be compared with each other. Hence, a model of a multi-state system having partially ordered state spaces is required in order to understand and solve practical reliability problems.

Mathematical studies about multi-state systems with totally ordered state spaces have been performed by many researchers (Barlow and Wu 1978, Griffith 1980, El-Neweihi et al. 1978, Natvig 1982, 2011, Ohi and Nishida 1983, 1984a, b, Ohi 2010). Huang et al. (2003) have extended a binary state consecutive k-out-of-n system which is well observed in a real situation to the multi-state case. Natvig (2011), Lisnianski and Levitin (2003) and Lisnianski et al. (2010) have summarized the work performed so far, where examples of practical applications of multi-state systems are presented. Ohi (2015) has given general stochastic upper and lower bounds for system's stochastic performances via a modular decomposition, which are convenient for systems designers and analysts.

For examinations of multi-state systems having partially ordered states, Levitin (2013) has proposed a new model of a multi-state k-out-of-n system called a multi-state vector-k-out-of-n system, of which state spaces are assumed to be a sub-set of the n-dimensional Euclid space. Yu et al. (1994), emphasizing the case that some states are possibly not to be ordered, have proposed a model of multi-state system having partially ordered state spaces. Ohi (2012, 2013, 2015, 2016a, b) has been trying to build up a general theoretical frame-work of a multi-state system. Ohi (2013) has given an existence theory of series and parallel systems and a series-parallel decomposition of a multi-state system, when the state spaces are lattice sets. Ohi (2012) has also given a characterization of a module by φ-equivalent relation under the lattice set assumption on the state spaces. Furthermore, Ohi (2015) has shown upper and lower bounds of $P\{\varphi \geq s\}$ via a modular decomposition. These bounds have been extended to $P(\varphi^{-1}(A))$ for an increasing set A by Ohi (2016b). In these works, the probability P on the product set of the state spaces of the components is assumed to be associated (Ohi 1989).

Based on these works, this chapter gives structural and probabilistic examinations of k-out-of-n:G system and a solution for a practical problem: How to construct an optimal structure of a safety monitoring system which is modelled as a Barlow and Wu (1978) multi-state system. The examination of the multi-state case under the entropy criterion assumption is newly given in this article.

The contemporary society is supported by many large-scale and complex systems, such as nuclear power plants, electric power networks, information networks, chemical plants, natural gas and oil pipe lines, and so on. Ensuring the security and safety of these systems is an important issue. Many researchers, utilizing the notions of reliability systems, have been studying the optimal structure of safety monitoring systems composed of many sensors (Ansel 1982, Hagihara et al. 1986a, Hagihara et al. 1986b, Inoue et al. 1982, Kohda et al. 1981, Nakajima 1987, Phillips 1976, 1980, Ohi and Suzuki 2000, Tanaka 1995).

In Section 3, we introduce a general definition of a multi-state system, where a binary state system is defined to be a special type of multi-state system. Though some researchers have already been studying the multi-state systems in the case of partially ordered state spaces, as mentioned above, in this chapter we suppose them to be totally ordered sets for the simplicity of our discussion. We may not generally adopt the max and min formulae for the definitions of the series and parallel systems, respectively. Definitions of these special systems are given in Section 5.

In Section 4, we explain the basic notions of binary state systems and show that the reliability function of a system may be given as a convex combination of k-out-of-n:G ($k = 1, ..., n$) systems when the components are stochastically identical and independent (Birnbaum et al. 1961). Using these facts, we show that an optimal structure of a safety monitoring system which is modelled by a binary state system is a k-out-of-n:G system under both cost and entropy criteria for the optimality.

In Section 5, using the general definition of the systems introduced in Section 3, we define k-out-of-n:G system. Furthermore, we show that the two kinds of relevant properties of Definitions 3.2 and 3.3 in Section 3 simultaneously restrict the cardinal numbers of the state spaces to be the same and then min and max formulae are logically justified. After these examinations, we focus on the multi-state systems by Barlow and Wu (1978) and show that the probability law of the system may be given by the convex combination of those of k-out-of-n:G ($k = 1, ..., n$) systems. We examine the optimal structure of a safety monitoring system modelled by Barlow and Wu's multi-state system and, using the convex combination, we show that a k-out-of-n:G system is optimal in both cases of cost and entropy criteria.

Sections 1 and 2 cover Introduction and Notations, respectively, and this chapter is closed with Section 6 for concluding remarks.

2. Notations

We use the following notations for a system (Ω_C, S, φ), for which the precise definition is given by Definition 3.1 in Section 3. $C = \{1, ..., n\}$, $\Omega_C = \Pi_{i \in C} \Omega_i$, Ω_i ($i \in C$) and S are finite totally ordered sets, and φ is a suijection from Ω_C to S.

1. For a subset $A \subseteq C$, $\Omega_A = \Pi_{i \in A} \Omega_i$
2. For $i \in C$ and $k \in \Omega_i$,

$$(\cdot_i, x) = (x_1, ..., x_{i-1}, \cdot, x_{i+1}, ..., x_n),$$

$$(k_i, x) = (x_1, ..., x_{i-1}, k, x_{i+1}, ..., x_n).$$

3. Letting $\left\{ B_j \mid 1 \leq j \leq m \right\}$ be a partition of a subset $A \subseteq C$, for

$x_j \in \prod_{i \in Bj} \Omega_i \, (1 \le j \le m)$, a state vector $x \in \Omega_A$ such that $P_{Bj} x = x_j \, (1 \le j \le m)$ is written as $x = (x_1, ..., x_m)$. Then for every $x \in \Omega_A \, (A \subseteq C)$, $x = (x^{B1}, ..., x^{Bm})$, where $x^{Bj} = P_{Bj} x \, (j = 1, ..., m)$ and P_{Bj} is the projection mapping from Ω_A to Ω_{Bj}.

4. When $\Omega_i = \{0, ..., N\}$, $i \in C$, $s \in \Omega_C$ denotes $s = (s, ..., s)$. For $x = (x_1 ..., x_n) \in \Omega_C$, $N - x = (N - x_1, ..., N - x_n)$.

5. Every order is commonly denoted by the symbol $\le$. $a \le b$ means b is greater than or equal to a. $a < b$ means $a \le b$ and $a \ne b$. Then, for $x, y \in \Omega_C$, $x \le y$ means $\forall i \in C$, $x_i \le y_i$, and $x < y$ means $\forall i \in C$, $x_i \le y_i$ and $\exists j \in C$, $x_j \ne y_j$.

6. For a subset A of S, $\varphi^{-1}(A)$ is the inverse image of A with respect to φ, i.e., $\varphi^{-1}(A) = \{x \mid \varphi(x) \in A\}$. When the bracket is cumbersome, it is dropped as $\varphi^{-1}A$ without any confusion.

7. For a subset $D \subseteq C$, φ^D is a structure function from Ω_D to S.

8. For an element x of an ordered set W, intervals $[x, \rightarrow)$ and $(\leftarrow, x]$ on W are defined as
$$[x, \rightarrow) = \{y \in W \mid x \le y\}, \, (\leftarrow, x] = \{y \in W \mid y \le x\}.$$
$MI(W)$ and $MA(W)$ denote the set of all the minimal and maximal elements of the ordered set W, respectively. An element $x \in W$ is called minimal (maximal), when there is no element $y \in W$ such that $y < x$ ($x < y$).

9. For sets A and B, $A \backslash B = \{x \mid x \in A, x \notin B\}$ is the difference between the sets of A and B. $\#A$ denotes the cardinal number of the set A.

10. $\prod$ and $\coprod$ means the product and coproduct of real numbers $x_i (1 \le i \le n)$, i.e., $\prod_{i=1}^{n} x_i = x_1 ... x_n$ and $\coprod_{i=1}^{n} x_i = 1 - \prod_{i=1}^{n} (1 - x_i)$.

3. General definition of a system

Definition 3.1. A multi-state system composed of n components is defined as a triplet (Ω_C, S, φ) satisfying the following conditions.

(i) $C = \{1, ... , n\}$ is a set of components.

(ii) $\Omega_i (i \in C)$ and S are finite totally ordered sets, denoting the state space of the i-th component and the system, respectively.

(iii) $\Omega_C = \prod_{i \in C} \Omega_i$ is the product ordered set of Ω_i ($i \in C$). Each element $x = (x_1, ..., x_n) \in \Omega_C$, where $x_i \in \Omega_i$, is called a state vector.

(iv) φ is a surjection from Ω_C to S, called a structure function, reflects an inner structure of the system, in other words, how the components are logically connected to determine the state of the system. The multi-state system is simply called a system φ, when there is no confusion.

When $\Omega_i = S = \{0,1\}$, $i \in C$, the system is called a binary state system.

Definition 3.2. (Definition of relevant property) (1) The component i of the system φ is called relevant when the following condition is satisfied.

$$\forall k, \forall l \in \Omega_i \text{ such that } k \neq l, \exists (\cdot_i, x) \in \Omega_{C \setminus \{i\}}, \varphi(k_i, x) \neq \varphi(l_i, x).$$

(2) When every component is relevant, the system is called relevant.

The relevant property is not a strong condition on the system (Ω_C, S, φ), since when a component i is not relevant, then we have

$$\exists k, \exists l \in \Omega_i \text{ such that } k \neq l, \forall (\cdot_i, x) \in \Omega_{C \setminus \{i\}}, \varphi(k_i, x) = \varphi(l_i, x).$$

which denotes that the states k and I of the component i contribute to the system's state entirely the same way, and so may be merged into one state.

The following definition gives us another notion of relevance from the system's state point of view.

Definition 3.3. (Definition of s-relevant property) (1) The component i of the system φ is called s-relevant, when for every s and t of S such that $s \neq t$, there exist (k_i, x) and (l_i, x) of Ω_C such that $\varphi(k_i, x) = s$ and $\varphi(l_i, x) = t$

(2) When every component is s-relevant, the system is called s-relevant.

The relevant property indicates that every different two states of a component differently contribute to the system's operation, however the contributed states of the system are not assigned. On the other hand, s-relevant property means that the component contributes differently to the arbitrary assigned different two states of the system.

Definition 3.4. (Definition of monotone property) A system is called monotone or increasing when the structure function is increasing, i.e., for $x, y \in \Omega_C, x \leq y$ implies $\varphi(x) \leq \varphi(y)$.

The monotone property means that the deterioration of the components does not improve the state of the system. Both of the monotone and relevant properties are considered to be practically natural for reliability systems and then we usually assume these conditions on the systems for their algebraic and probabilistic examinations.

Definition 3.5. A system φ is called a coherent system when the system is relevant and monotone.

Definition 3.6. The dual system of a system (Ω_C, S, φ) is defined to be a system $\left(\Omega_C^{Du}, S^{Du}, \varphi^{Du} \right)$ as

(i) Ω_C^{Du} is the dual ordered set of Ω_C, i.e., the product ordered set of $\Omega_i^{Du} (i \in C)$, where Ω_i^{Du} is the dual ordered set of Ω_i.

(ii) S^{Du} is the dual ordered set of S.

(iii) φ^{Du} is a mapping from Ω_C^{Du} to S^{Du} defined as

$$x \in \Omega_C^{Du}, \varphi^{Du}(x) = \varphi(x).$$

Denoting the dual order commonly by $\leq^{Du}$,

$$x, y \in \Omega_C^{Du}, x \leq^{Du} y \Rightarrow \varphi^{Du}(x) \leq^{Du} \varphi^{Du}(y),$$

when the system φ is monotone, since

$$x \leq^{Du} y \Rightarrow y \leq x \Rightarrow \varphi(y) \leq \varphi(x) \Rightarrow \varphi(x) \leq^{Du} \varphi(y) \Rightarrow \varphi^{Du}(x) \leq^{Du} \varphi^{Du}(y).$$

We easily arrive at the following theorem.

Theorem 3.1. (1) When a system is monotone, the dual system is monotone, and vice versa.

(2) When a system is relevant (s-relevant), the dual system is relevant (s-relevant), and vice versa.

4. Binary state *k*-out-of-*n*:G system

In this section, we focus on binary state systems and examine binary state k-out-of-n: G ($k = 1, ..., n$) systems, whose application to the structural optimality problem of safety monitoring system is also shown.

4.1 A binary state coherent system

4.1.1 A binary state coherent system

The state spaces of a binary state system (Ω_C, S, φ) are assumed to be $\Omega_i = S = \{0, 1\}$ ($i \in C$), which mean that we only consider the operating and failure states and not the intermediate states between them. In this case, the two kinds of relevant properties are coincident with each other and are defined for the component $i \in C$ as the following.

$$\exists (\cdot_i, x), \in \Omega_{C\setminus\{i\}}, \varphi(0_i, x) \neq \varphi (1_i, x).$$

If the component is not relevant, $\varphi(0_i, x) = \varphi(1_i, x)$ holds for every $(\cdot_i, x)$, and then the states 0 and 1 contribute equivalently to the system's operation. The states 0 and 1 may then be merged into one state, and furthermore, we may eliminate the component i from the system, which means that the irrelevant component i is dummy.

The dual system of a binary state system (Ω_C, S, φ) may be written as the system $(\Omega_C, S, \varphi^{Du})$ defined by

$$x \in \Omega_C, \varphi^{Du}(x) = 1 - \varphi (1-x) = 1 - \varphi(1 - x_1, ... , 1 - x_n). \tag{1}$$

Example 4.1. (1) A system φ is called a series system, when the structure function is defined as the following.

$$x \in \Omega_C, \varphi(x) = \begin{cases} 1, & \#\{i \mid x_i = 1\} = n \\ 0, & \#\{i \mid x_i = 0\} \geq 1. \end{cases}$$

A series system needs all the components to function for the system's operation. It is easily proved that a system is a series if and only if the following condition is satisfied.

$$\forall x \in \Omega_C, \ \forall y \in \Omega_C, \varphi(x) \wedge \varphi(y) = \varphi(x \wedge y),$$

which means that the serialization at the component level and the system level are the same.

(2) A system φ is called a parallel system when the structure function is defined as the following:

$$x \in \Omega_C, \varphi(x) = \begin{cases} 1, & \#\{i \mid x_i = 1\} \geq 1, \\ 0, & \#\{i \mid x_i = 0\} = n. \end{cases}$$

A parallel system functions when at least one component functions and the dual system is a series system. We have the similar necessary and sufficient conditions for a system to be parallel. The dual system of the series system is the parallel system, and vice versa.

These series and parallel systems are special cases of the k-out-of-n:G system of which definition is given in Section 4.2.

4.1.2 Minimal path and cut vectors

For a system (Ω_C, S, φ), noticing that the product set Ω_C is a finite ordered set with the product order defined on it, we have minimal and maximal state vectors for every sub-set of Ω_C.

The minimal state vectors of $\varphi^{-1}(1)$ are called the minimal path vectors of the system and the maximal state vectors of $\varphi^{-1}(0)$ are called the minimal cut vectors of the system.

Defining $C_k(x)$ for a state vector $x \in \Omega_C$ as

$$C_k(x) = \{i \mid x_i = k\}, k = 0, 1,$$

for $a \in MI(\varphi^{-1}(1))$, $C_1(a)$ is called a minimal path set, and $C_0(b)$ is called a minimal cut set for a state vector $b \in MA(\varphi^{-1}(0))$.

Example 4.2. (1) When a system φ is a series system, then $MI(\varphi^{-1}(1))$ = $\{(1, ..., 1)\}$ which means that the inverse set $\varphi^{-1}(1)$ has the minimum state vector. This property is a necessary and sufficient condition for the system to be series.

(2) When a system φ is a parallel system, then $\varphi^{-1}(0)$ has the maximum state vector $(0, ..., 0)$ and then $MA(\varphi^{-1}(0)) = \{(0, ... , 0)\}$, which is also a necessary and sufficient condition for the system to be parallel.

The properties shown in Example 4.2 tell us how to define multi-state series and parallel systems. See Section 5.1.2.

The following relation is easily verified. For every state vector $x \in \Omega_C$,

$$\varphi(x) = 1 \Leftrightarrow \exists a \in MI(\varphi^{-1}(1)), x \geq a,$$

$$\varphi(x) = 0 \Leftrightarrow \exists b \in MI(\varphi^{-1}(0)), x \leqq b.$$

Hence, we have the max – min formulae of the structure function φ as

$$x \in \Omega_C, \varphi(x) = \max_{a \in MI(\varphi^{-1}(1))} \min_{i \in C_1(a)} x_i \qquad (2)$$

$$= \min_{b \in MA(\varphi^{-1}(0))} \max_{i \in C_0(b)} x_i \qquad (3)$$

These formulae denote that the structure function may be decomposed into a family of series structure functions or parallel structure functions. Using these formulae, writing the indicator function of $k \in \{0, 1\}$ as I_k, $k = 0, 1$, we have a Boolean representation of the structure function as

$$x \in \Omega_C, \varphi(x) = \coprod_{a \in MI(\varphi^{-1}(1))} \prod_{i \in C_1(a)} I_1(x_i) \qquad (4)$$

$$= \prod_{b \in MA(\varphi^{-1}(0))} \coprod_{i \in C_0(b)} I_0(x_i) \qquad (5)$$

which ensures that we have a multinomial representation of the reliability function when the components are stochastically independent.
Noticing

$$a, a' \in MI(\varphi^{-1}(1)) \text{ such that } a \neq a', C_1(a)\backslash C_1(a') \neq \phi,$$

$$\bigcup_{a \in MI(\varphi^{-1}(1))} C_1(a) = C,$$

we have uniquely a coherent structure function φ_D for a given family of subsets of C, $\mathcal{D} = \{D_1, \dots, D_p\}$, such that $D_i \backslash D_j \neq \phi$, $i \neq j$ and $\bigcup_{j=1}^{p} D_j = C$. $\mathcal{D}$ must be the family of the minimal path sets of the structure function φ_D, which is defined as

$$x \in \Omega_C, \varphi_D(x) = \max_{1 \leq j \leq p} \min_{i \in D_j} x_i.$$

It is easily shown that the minimal path and cut sets of the dual system are the minimal cut and path sets of the original system, respectively.

4.1.3 Definition of $A_j(\varphi)$ and its properties

For a binary state system (Ω_C, S, φ) with n components, we define $A_j(\varphi)$ by

$$A_j(\varphi) = \#\{x \in \Omega_C \mid \varphi(x) = 1, \#C_1(x) = j\}, j = 0, 1, \dots, n, \qquad (6)$$

which is used to express the reliability function as a convex combination of the reliability functions of k-out-of-n:G ($k = 1, \dots, n$) systems. When there is no confusion, we simply write A_j for $A_j(\varphi)$. Apparently, $A_0 = 0$ and $A_n = 1$.

Theorem 4.1. We have the following inequality.

$$(j + 1)A_{j+1} \geq (n - j)A_j, j = 0, 1, \dots, n - 1.$$

Proof. We temporarily use a symbol $V_j(\varphi)$ to denote

$$V_j(\varphi) = \{x \in \Omega_C \mid \varphi(x) = 1, \# C_1(x) = j\}$$

of which the cardinal number is $A_j(\varphi)$. $V_j(\varphi)$ is simply written as V_j. Noticing

for $y \in V_{j+1}$, $\#\{x \in \Omega_C \mid y \geq x, \#C_1(x) = j\} = j + 1$,

for $x \in V_j$, $\#\{y \in \Omega_C \mid y \geq x, \#C_1(y) = j + 1\} = n - j$,

and the following inclusion relation by the monotonic property of φ,

$$\bigcup_{y \in V_{j+1}} \{x \in \Omega_C \mid y \geq x, \# C_1(x) = j\} \times \{y\}$$

$$\supseteq \bigcup_{x \in V_j} \{x\} \times \{y \in \Omega_C \mid y \geq x, \# C_1(y) = j+1\},$$

we have the next inequality by simple calculation of the cardinal numbers

$$\sum_{y \in V_{j+1}} \#\{x \in \Omega_C \mid y \geq x, \# C_1(x) = j\}$$

$$\geq \sum_{x \in V_j} \#\{y \in \Omega_C \mid y \geq x, \# C_1(y) = j + 1\}$$

Hence, the aimed inequality holds.

By Theorem 3.1, we have the following corollary.

Corollary 4.1. For a system (Ω_C, S, φ) having n components, the next inequalities hold.

$$\frac{A_j(\varphi)}{\binom{n}{j}} \leq \frac{A_{j+1}(\varphi)}{\binom{n}{j+1}}, j = 0, \ldots, n - 1$$

Example 4.3. (Bridge system) The bridge system composed of five components is defined by minimal path sets as

$$\{1,4\}, \{2,5\}, \{1,3,5\}, \{2,3,4\},$$

and then V_j, $j = 1, 2, 3, 4, 5$ are given as

$V_1 = \phi$,

$V_2 = \{\{1,4\},\{2,5\}\}$,

$V_3 = \{\{1,2,4\}, \{1,3,4\}, \{1,4,5\}, \{1,2,5\}, \{2,3,5\}, \{2,4,5\}, \{1,3,5\}, \{2,3,4\}\}$,

$V_4 = \{\{1,2,3,5\},\{1,2,4,5\},\{1,3,4,5\},\{1,2,3,4\},\{2,3,4,5\}\}$,

$V_5 = \{\{1,2,3,4,5\}\}$,

and then we have, by taking the cardinal numbers of the above sets,

$$A_1 = 0, A_2 = 2, A_3 = 8, A_4 = 5, A_5 = 1$$

Furthermore, we may verify the assertion of Corollary 3.1 with this example.

$$\frac{A_1}{\binom{5}{1}} = 0, \ \frac{A_2}{\binom{5}{2}} = \frac{1}{5}, \ \frac{A_3}{\binom{5}{3}} = \frac{4}{5}, \ \frac{A_4}{\binom{5}{4}} = 1, \ \frac{A_5}{\binom{5}{5}} = 1$$

4.2 *k*-out-of-*n*:G system

A system (Ω_C, S, φ) having n components is called k-out-of-n:G system, when the following conditions hold.

$$x \in \Omega_C, \varphi(x) = \begin{cases} 1, & \#C_1(x) \geq k, \\ 0, & \#C_1(x) \leq k-1, \end{cases}$$

which denotes that the system needs at least k operating components for its operation. The structure function of k-out-of-n:G system is written as $\varphi_{k.n.G}$. Apparently, the n-out-of-n:G and 1-out-of-n:G systems are, respectively, the series and parallel systems.

The minimal path and cut vectors of k-out-of-n:G system are

$$MI\left(\varphi_{k,n,G}^{-1}(1)\right) = \{x | \#C_1(x) = k\},$$

$$MA\left(\varphi_{k,n,G}^{-1}(0)\right) = \{x | \#C_0(x) = n-k+1\},$$

and then the minimal path and cut sets are respectively given as

$$\{C_1(x) | \#C_1(x) = k\} = \{\{i_1,...,i_k\} | 1 \leq i_1 < \cdots < i_k \leq n\},$$

$$\{C_0(x) | \#C_0(x) = n-k+1\} = \{\{i_1,...,i_{n-k+1}\} | 1 \leq i_1 < \cdots < i_{n-k+1} \leq n\}.$$

$A_j\left(\varphi_{k,n,G}\right), j = 0,\cdots,n$, are easily obtained as

$$A_j\left(\varphi_{k,n,G}\right) = \begin{cases} \binom{n}{j}, & j \geq k, \\ 0, & j \leq k-1. \end{cases}$$

The k-out-of-n:F system is defined to have the following structure function $\varphi_{k,n,F}$.

$$x \in \Omega_C, \varphi_{k,n,F}(x) = \begin{cases} 1, & \#C_0(x) \leq k-1, \\ 0, & \#C_0(x) \geq k. \end{cases}$$

It is easily shown that the dual system of the k-out-of-n:G system is the $n - k + 1$-out-of-n:G system. Hence, the minimal path and cut sets and vectors of this system are easily given from those of the k-out-of-n:G system. For example, the minimal path sets of $n - k + 1$-out-of-n:G system are the minimal cut sets of the k-out-of-n:G system.

4.2.1 An alternative definition of k-out-of-n:G system

For k-out-of-n:G system $(\Omega_C, S, \varphi_{k,n,G})$, each subset $D \subseteq C$ such that $\#D = k$ is a minima path set, then denoting the series system defined by the minimal path set as $\varphi^D : \Omega_D \to S$, the structure function $\varphi_{k,n,G}$ is expressed as

$$x \in \Omega_C, \varphi_{k,n,G}(x) = \max_{D \subseteq C, \#D=k} \varphi^D(x).$$

This decomposition shows us an alternative definition of k-out-of-n:G system as the following.

Definition 4.1. (An alternative definition of k-out-of-n:G system) A system (Ω_C, S, φ) is called k-out-of-n:G system when the structure function φ is decomposed into a family of series systems (Ω_D, S, φ^D) $(D \subseteq C, \#D = k)$ as

$$x \in \Omega_C, \varphi(x) = \max_{D \subseteq C, \#D=k} \varphi^D(x^D).$$

The alternative Definition 4.1 does not depend on the cardinal number of the state spaces and may be used to define a general multi-state k-out-of-n:G in Section 5.

4.3 Reliability function

4.3.1 Definition of reliability function

Let (Ω_C, S, φ) be an n-component coherent system having a probability P on Ω_C, which denotes joint stochastic performances of the components. When the components are stochastically independent, the probability P is expressed by $(p_1, ..., p_n)$, where p_i is the probability that the component i is (normal) operating and the failure probability of the component is $1 - p_i$.

The reliability function h_φ is a mapping, by which a probability P corresponds to the image probability $\varphi \circ P$ on S,

$$h_\varphi(P) = \varphi \circ P,$$

and for $X \subseteq S$, $\varphi \circ P(X) = P(\varphi^{-1}(X))$.

For a binary state system, the state space of the system is $S = \{0,1\}$, and then the probability $h_\varphi(P)$ is determined by assigning the probability that the system is normal (functioning), i.e., the probability of the state 1. Then, we may equivalently define the reliability function as a mapping

$$h_\varphi(P) = P\left(\varphi^{-1}(1)\right).$$

When the components are stochastically independent, the probability P is given by $P = (p_1, ..., p_n)$ and then $h_\varphi(P)$ is a multinomial function of $(p_1, ..., p_n)$ by taking the expectation of the Boolean expressions of the structure function, (4) and (5), in which case we write $h_\varphi(p_i, ..., p_n)$ or $h_\varphi(p)$. Furthermore, when the components are stochastically identical with operating probability p of each component, the reliability function is simply written as $h_\varphi(p)$.

Example 4.4. (1) For a series system, $\varphi^{-1}(1) = \{(1, ..., 1)\}$ and then

$$h_\varphi(P) = P\{(1, ..., 1)\}.$$

When the components are stochastically independent,

$$h_\varphi(p) = \prod_{i=1}^{n} p_i.$$

(2) For a parallel system, noticing $\varphi^{-1}(0) = \{(0, ..., 0)\}$, then we have

$$h_\varphi(P) = 1 - P\{(0, ..., 0)\}$$

and in the independent case

$$h_\varphi(p) = 1 - \prod_{i=1}^{n}(1 - p_i) = \coprod_{i=1}^{n} p_i.$$

(3) The reliability function of the k-out-of-n:G system is somewhat complicated. $V_j, j = 1, ..., n$, of the k-out-of-n:G system are given as the following:

$$V_j = \begin{cases} \phi, & j \leq k-1, \\ \left\{\left(1^D, 0^{C/D}\right) | D \subseteq C, \#D = j\right\}, & j \geq k. \end{cases}$$

Then

$$h_{\varphi_{k,n,G}}(P) = \sum_{j=k}^{n} P(V_j) = \sum_{j=k}^{n} \sum_{x \in V_j} P(x)$$

It is easy for every structure function φ and probability P on Ω_c,

$$P(1) = \varphi_{n,n,G} \circ P(1) \leq \varphi \circ P(1) \leq \varphi_{1,n,G} \circ P(1) = 1 - P(0), \tag{7}$$

which denotes that the reliabilities of the series and parallel systems are, respectively, the minimum and maximum among the systems composed of n components.

Even if the components are stochastically independent, the derivation of the reliability function of a binary state system is complex, as shown in Example 4.4 (3). Typical methods are inclusion-exclusion principle, taking expectation of Boolean expression of the structure function. Some useful stochastic bounds of the system's reliability are given (Barlow and Proschan 1975). These methods and bounds are based on the series or parallel decomposition of the structure functions (2)(3).

When the components are stochastically identical and independent, the reliability function is very simply expressed as a convex combination of those of the k-out-of-n:G systems, $k = 1,..., n$.

4.3.2 Reliability function in the case of i.i.d. components

When the components of a system (Ω_c, S, φ) are stochastically independent and identical, then the probability P on Ω_c is given as $P = (\underbrace{p,...,p}_{n})$, where

p is the probability that each component is functioning. Using the symbol V_j, given in the proof of Theorem 4.1, it is easy to have

$$\varphi^{-1}(1) = \bigcup_{j=1}^{n} V_j,$$

then the reliability function in this case of identical and independent components is given as

$$h_\varphi(p) = \sum_{j=1}^{n} A_j(\varphi) p^j (1-p)^{n-j}$$

since, noticing that for each state vector $x \in V_j$, $P(x) = p^j(1-p)^{n-j}$, and V_j, $j = 1, ..., n$, are mutually disjoint,

$$h_\varphi(p) = P(\varphi^{-1})(1) = P\left(\bigcup_{j=1}^{n} V_j\right) = \sum_{j=1}^{n} P(V_j) = \sum_{j=1}^{n} A_j p^j (1-p)^{n-j}.$$

Hence by Corollary 4.1, we have the next Theorem 4.2.

Theorem 4.2. For a coherent system (Ω_c, S, y) with probability $P = (p, ..., p)$ on Ω_c, the reliability function $h\varphi(p)$ is given as the following convex combination of those of k-out-of-n: G, $k = 1, ..., n$ systems.

$$h_\varphi(p) = \sum_{k=1}^{n} \alpha_k h_{\varphi k,n,G}(p), \quad \alpha_k = \frac{A_k}{\binom{n}{k}} - \frac{A_{k-1}}{\binom{n}{k-1}},$$

where $\sum_{k=1}^{n} \alpha_k = 1, \alpha_k \geq 0$.

Proof. We may easily verify the following calculation.

$$h_\varphi(p) = \sum_{j=1}^{n} \frac{A_j}{\binom{n}{j}} \binom{n}{j} p^j (1-p)^{n-j}$$

$$= \sum_{j=1}^{n} \left(\sum_{k=1}^{j} \alpha_k \right) \binom{n}{j} p^j (1-p)^{n-j}$$

$$= \sum_{k=1}^{n} \alpha_k \sum_{j=k}^{n} \binom{n}{j} p^j (1-p)^{n-j}$$

$$= \sum_{k=1}^{n} \alpha_k h_{\varphi_{k,n,G}}(p).$$

4.3.3 Entropy of a binary coherent system

For a probability P on Ω_C, the entropy of the system,

$$H_\varphi(P) = -P(\varphi^{-1}(1)) \log P (\varphi^{-1}(1)) - P(\varphi^{-1}(0)) \log P (\varphi^{-1}(0)),$$

denotes the ambiguity level of the signals outputted by the system. Letting $f(x) = -x \log x - (1-x) \log (1-x)$, $x \in [0,1]$, the entropy is simply written as $H_\varphi(P) = f(P(\varphi^{-1}(1)))$. The function f has typical properties, it is increasing and decreasing on $[0, 1/2]$ and $[1/2,1]$, respectively, and concave on $[0,1]$ and symmetric at $x = 1/2$.

The dual probability P^{Du} of a probability P on Ω_C is defined to satisfy

$$x \in \Omega_C, \ P^{Du}(x) = P(1-x).$$

Then, noticing the definition of the dual system (1), the following theorem is clear.

Theorem 4.3. For a binary coherent system (Ω_C, S, φ) and a probability P on Ω_C,

$$H_\varphi(P) = H_{\varphi Du} (P^{Du}).$$

Theorem 4.4. For a binary coherent system (Ω_C, S, φ) and a probability P on Ω_C, we have the following relations.

$$P(0) \le P(1) \Rightarrow H_\varphi(P) \ge H_{\varphi_{1, n,G}}(P),$$

$$P(0) \ge P(1) \Rightarrow H_\varphi(P) \ge H_{\varphi_{n, n,G}}(P).$$

For the proof, refer to Ohi and Suzuki (2000).

Theorem 4.4 does not necessarily require the stochastic independence of the components and shows us that the minimum entropy of the system's output is given by the series or the parallel system depending on whether $P(1) \geq P(0)$ holds or not.

Corollary 4.2. When the components are stochastically identical and independent, i.e., the probability P on Ω_C is given as $P = (\underbrace{p, \cdots, p}_{n})$, then we have the following relations.

$$p \geq 1/2 \Rightarrow H_{\varphi}(P) \geq H_{\varphi_{1,n,G}}(P),$$

$$p \leq 1/2 \Rightarrow H_{\varphi}(P) \geq H_{\varphi_{n,n,G}}(P).$$

Proof. Noticing $P(0) = (1 - p)^n$ and $P(1) = p^n$, Corollary is clear from Theorem 4.4.

4.4 Optimal structure of a safety monitoring system

A safety monitoring system composed of n sensors is formulated as a binary state system (Ω_C, S, φ) defined by Definition 3.1, where 0 and 1 mean for each sensor not to give and to give an alarm, respectively. For example, a state vector of three components (0,1,1) denotes that the sensor 1 gives no signal and the sensors 2 and 3 give signals. For each combination of the actions of the sensors, $x \in \Omega_C$, $\varphi(x)$ shows us whether the safety monitoring system gives us an alarm or not. In other words, the structure function φ indicates a decision maker for giving an alarm based on the observations of the environment by the sensors.

In this section we examine the structure function and show one of the k-out-of-n:G ($k = 1, ..., n$) systems to be optimal in the sense of minimizing the cost or entropy, when the sensors are stochastically identical and independent.

4.4.1 Formulation

Suppose E to be the state space of the environment monitored by a safety monitoring system (Ω_C, S, φ). We do not set an assumption about the cardinal number of the state space E, but which is endowed with an appropriate σ-algebra $\mathcal{E}$. Ω_i, $i \in C$ and S are finite sets, then we consider power sets as their σ-algebras, $\mathcal{P}(\Omega_i)$, $\mathcal{P}(S)$, respectively. The power set $\mathcal{P}(\Omega_C)$ is of course the product σ-algebra of $\mathcal{P}(\Omega_i)$, $i \in C$. We assume U to be a probability on $(E, \mathcal{E})$, which denotes a probability that an environmental event occurs, and $P(e, \cdot)$ to be a conditional probability on Ω_C when the state of the environment is e, in other words, $P(\cdot, \cdot)$ is mathematically a transition probability from $(E, \mathcal{E})$ to $(\Omega_C, \mathcal{P}(\Omega_C)')$. For $e \in E$ and $x \in \Omega_C$, $P(e, x)$ practically denotes a probability that the sensors jointly take the action x when the environmental state is e.

Depending on the structure function φ, we may define a transition probability $Q_\varphi(\cdot,\cdot)$ from (E, ε) to $(S, \mathcal{P}(S))$ as

$$e \in E, s \in S, Q_\varphi(e, s) = P(e, \varphi^{-1}(s)),$$

meaning the probability that the safety system having the structure function φ takes an action $s \in S$ when the environment is in the state e.

4.4.2 Optimal structure minimizing cost

A cost function R defined on $E \times S$ gives us the cost incurred when the environment is $e \in E$ and the action $x \in \Omega_C$ is taken. The expected cost is given by

$$E[R] = \int_{e \in E} \int_{s \in S} R(e,s) Q_\varphi(e,ds) U(de)$$

$$= \int_{e \in E} \left\{ R(e,0) Q_\varphi(e,0) + R(e,1) Q_\varphi(e,1) \right\} U(de)$$

$$= \int_{e \in E} R(e,0) U(de) + \int_{e \in E} (R(e,1) - R(e,0)) Q_\varphi(e,1) U(de),$$

when the safety system's structure function is φ.

When the sensors are stochastically independent and for each $e \in E$, $P(e,\cdot)$ is the product probability of $P_i(e,\cdot)$, $i \in C$, which is the transition probability from (E, ε) to $(\Omega_i \mathcal{P}(\Omega_i))$. Using the reliability function h_φ of the system φ, we have

$$E[R] = \int_{e \in E} R(e,0) U(de)$$

$$+ \int_{e \in E} (R(e,1) - R(e,0)) h_\varphi (P_1(e,1), \cdots, P_n(e,1)) U(de)$$

Furthermore, if the sensors are identical, i.e.,

$$e \in E, P_i(e, 1) = P(e, 1), i = 1, \ldots, n,$$

then the following formula is given by Theorem 4.2,

$$E[R] = \int_{e \in E} R(e,0) U(de)$$

$$+ \sum_{k=1}^{n} \alpha_k \int_{e \in E} (R(e,1) - R(e,0)) h_{k,n,G} (P(e,1)) U(de),$$

hence, we have the following theorem.

Theorem 4.5. When the sensors are stochastically independent and identical, then the optimal system minimizing the cost is given by a k^*

-out-of-n: G system, where k^* attains the minimum

$$\min_{1\leq k\leq n} \int_{e\in E} \big(R(e,1)-R(e,0)\big)h_{k,n,G}\big(P(e,1)\big)U(de).$$

4.4.3 Optimal structure minimizing conditional entropy

For the safety monitoring system φ, the conditional entropy of the system with respect to the monitored system, is given as

$$H(\varphi\,|\,E) = \int_{e\in E} H_\varphi\big(P(e,\cdot)\big)U(de)$$

$$= \int_{e\in E}\Big\{-P\big(e,\varphi^{-1}(1)\big)\log P\big(e,\,\varphi^{-1}(1)\big)$$
$$-P\big(e,\varphi^{-1}(0)\big)\log P\big(e,\,\varphi^{-1}(0)\big)\Big\}U(de).$$

Then the following theorem holds, which asserts that the system's structure minimizing the conditional entropy level is a k-out-of-n:G system, when the sensors are stochastically identical and independent. We omit the proof which is easily given by the concavity of f.

Theorem 4.6. When the sensors are stochastically identical and independent, we have the following inequality.

$$H(\varphi\,|\,E) \geq \min_{1\leq k\leq n} H(\varphi_{k,n,G}\,|\,E)$$

5. Multi-state *k*-out-of-*n*:G system with finite totally ordered state spaces

In this section, we examine structural properties of multi-state k-out-of-n:G ($k = 1, \ldots , n$) systems and show that the optimal structure of a safety monitoring system is a k-out-of-n:G system under an assumption that the type of the system is restricted to be the Barlow and Wu system.

For a system (Ω_C, S, φ), the state spaces Ω_i ($i \in C$) and S are assumed to be finite totally ordered sets and not necessarily binary. We may assume $\Omega_i = \{0,1, \ldots, N_i\}$, $S = \{0,1, \ldots, N\}$ without loss of generality. The components and the system are assumed to have intermediate deteriorating states between failure and normal states.

Here, we notice that we use the same symbol, for example, 1 for different components, but the state 1's of different components do not denote the same state, so we may not adopt the formula as $\min_{i\in C} x_i$ for the definition of a series system, since this formula requires us to compare the states of the components and determine the minimum of them. We need more essential definitions of series and parallel systems.

5.1 Multi-state system with finite totally ordered state spaces

5.1.1 Multi-state system

Similarly to the fact that a binary state system is uniquely determined by a set of minimal path vectors or of cut vectors, the structure function of an increasing system φ is uniquely determined by a family of minimal state vectors $\{MI(\varphi^{-1}[s, \rightarrow))\}_{s \in S}$ or equivalently by a family $\{MI(\varphi^{-1}(s))\}_{s \in S}$. Considering the dual order, the increasing system is similarly determined by a family of maximal state vectors $\{MA(\varphi^{-1}(\leftarrow,s))\}_{s \in S}$ or $\{MA(\varphi^{-1}(s))\}_{s \in S}$.

Theorem 5.1. For an increasing system (Ω_C, S, φ), the following equivalent relations hold.

$$(1) \quad \varphi(x) = s \Leftrightarrow \begin{cases} \exists a \in MI\left(\varphi^{-1}[s, \rightarrow)\right), x \geq a, \\ \forall t \text{ such that } t > s, \forall b \in MI\left(\varphi^{-1}[t, \rightarrow)\right), x \not\geq b, \end{cases} \tag{8}$$

$$\Leftrightarrow s = \max\{t \mid \exists a \in MI\ (\varphi^{-1}\ [t, \rightarrow)), x \geq a\},$$

$$(2) \quad \varphi(x) \geq s \Leftrightarrow \exists a \in MI(\varphi^{-1}[s, \rightarrow)), x \geq a. \tag{9}$$

Proof. The first and the third equivalent relations are clear. We suppose $\varphi(x) = s$ and denote $A = \{t \mid \exists a \in MI(\varphi^{-1}[t, \rightarrow)), a \leq x\}$. $s \in A$ follows from the assumption of $\varphi(x) = s$. For $t \in A$,

$$\exists a \in MI(\varphi^{-1}[t, \rightarrow)), a \leq x,$$

and then $t \leq \varphi(a) \leq \varphi(x) = s$. Hence, $t \leq s$, which means s is the maximum element of A.

Example 5.1. We consider an increasing system, composed of three components, of which state spaces are given as $\Omega_1 = \Omega_2 = \{0,1,2\}$, $\Omega_3 = \{0,1\}$, $S = \{0,1,2\}$. The structure function φ is defined by giving minimal state vectors as

$$MI(\varphi^{-1}[1, \rightarrow)) = \{(0,1,1), (1,1,0), (2,0,1)\},$$

$$MI(\varphi^{-1}[2, \rightarrow)) = \{(1,2,1), (2,1,0), (2,0,1)\}.$$

The following minimal state vectors are also clearly given:

$$MI(\varphi^{-1}(1)) = \{(0,1,1), (1,1,0)\},$$

$$MI(\varphi^{-1}(2)) = \{(1,2,1), (2,1,0), (2,0,1)\}.$$

5.1.2 Series and parallel systems

In this section, following Ohi (2016a), we present a rough sketch of discussion about series and parallel systems. Following the comments given in Example 4.2, we define series and parallel systems as follows:

Definition 5.1. (1) A system φ is called a series system, when for every $s \in S$, $\varphi^{-1}[s, \rightarrow)$ has the minimum element.

(2) A system φ is called a parallel system, when for every $s \in S$, $\varphi^{-1}(\leftarrow, s]$ has the maximum element.

In the binary case, the necessary and sufficient condition for a system to be a series (parallel) system is that the serialization (parallelization) at component and system levels are equivalent. For the multi-state series and parallel systems defined by Definition 5.1, the similar relations hold, i.e., we have the following equivalent relations. For every x and $y \in \Omega_C$,

$$\varphi\text{: series} \Leftrightarrow \varphi(x) \wedge \varphi(y) = \varphi(x \wedge y),$$

$$\varphi\text{: parallel} \Leftrightarrow \varphi(x) \vee \varphi(y) = \varphi(x \vee y).$$

which means the validity of our definition. $\wedge$ and $\vee$ denote the inf and sup, respectively. Definition 5.1 means that the series and parallel systems have some special structural properties, so we may not consider these structures for arbitrary given state spaces. Therefore, we have an existence theorem of series and parallel systems.

Theorem 5.2. For a series system (Ω_C, S, φ), we have the following inequality from the relevant property.

$$\max_{i \in C} \# \Omega_i \leq \# S.$$

Furthermore, if the system is s-relevant we have another inequality.

$$\# S \leq \min_{i \in C} \# \Omega_i.$$

Then, when the series system is s-relevant and relevant, from both of the above inequalities we have

$$\# S = \# \Omega_i, i \in C.$$

Hence, we may assume $S = \Omega_i = \{0, 1, \ldots, N\}$ $(i \in C)$, from which $MI(\varphi^{-1}(s)) = \{(s, \ldots, s)\}$ is given.

By the dual examination, the similar assertion holds for a parallel system.

From Theorem 5.2, we may logically derive descriptive formulae of s-relevant series and parallel systems' structure functions as

$$\varphi(x) = \min_{1 \leq i \leq n} x_i, \quad \varphi(x) = \max_{1 \leq i \leq n} x_i,$$

respectively. The precise examinations are shown in Ohi (1983, 2016a).

The equations (2) and (3) give us decompositions of a structure function of a binary coherent system into a family of series systems or parallel systems. Following the definitions of series and parallel systems given by Definition 5.1, we may similarly decompose a multi-state structure function into a family of series systems or parallel systems, the precise procedure may be seen in Ohi (2016a).

5.2 Multi-state *k*-out-of-*n*:G system

5.2.1 Definition and structure of k-out-of-n:G system

Referring to the alternative definition of binary state *k*-out-of-*n*:G system given by Definition 4.1 and using the notion of multi-state series and parallel systems given by Definition 5.1, we give a definition of *k*-out-of-*n*:G system as follows:

Definition 5.2. (1) A system (Ω_C, S, φ) is called *k*-out-of-*n*:G system, of which structure function is written as $\varphi_{k,n,G}$, when for every $D \subseteq C$ such that $\#D = k$, there exists a series system (Ω_D, S, φ^D) and

$$x \in \Omega_C, \varphi_{k,n,G}(x) = \max_{D \subseteq C, \#D=k} \varphi^D(x^D). \tag{10}$$

2. A system (Ω_C, S, φ) is called *k*-out-of-*n*:F system, of which structure function is written as $\varphi_{k,n,F}$, when for every $D \subseteq C$ such that $\#D = k$, there exists a parallel system (Ω_D, S, φ^D) and

$$x \in \Omega_C, \varphi_{k,n,F}(x) = \min_{D \subseteq C, \#D=k} \varphi^D(x^D). \tag{11}$$

Writing the minimal state vector of $(\varphi^D)^{-1}(s)$ as $m_s^D \in \Omega_D$, from (10), the minimal state vectors of $\varphi_{k,n,G}^{-1}[s, \rightarrow)$ is given as

$$s \in S, MI\left(\varphi_{k,n,G}^{-1}[s, \rightarrow)\right) = MI\left(\varphi_{k,n,G}^{-1}(s)\right) = \bigcup_{D \subseteq C, \#D=k} \left\{(m_s^D, 0^{C \backslash D})\right\}.$$

The maximal state vectors of the *k*-out-of-*n*:F system are similarly given.

All the systems in this chapter are assumed to be coherent, and then relevant. If φ^D $(D \subseteq C, \#D = k)$ are also s-relevant, by Theorem 5.2 about multi-state series and parallel systems, $\#S = \#\Omega_i$ $(i \in C)$ holds, so (10) (11) may be rewritten as

$$x \in \Omega_C, \varphi_{k,n,G}(x) = \max_{D \subseteq C, \#D=k} \min_{i \in D} x_i,$$

$$x \in \Omega_C, \varphi_{k,n,F}(x) = \min_{D \subseteq C, \#D=k} \max_{i \in D} x_i.$$

The minimal state vectors are given as

$$MI\left(\varphi_{k,n,G}^{-1}[s, \rightarrow)\right) = \left\{(s^D, 0^{C \backslash D}) \mid D \subseteq C, \#D = k\right\}, \tag{12}$$

$$MA\left(\varphi_{k,n,F}^{-1}(\leftarrow, s]\right) = \left\{(s^D, 0^{C \backslash D}) \mid D \subseteq C, \#D = k\right\}. \tag{13}$$

5.2.2 Dual system of k-out-of-n:G system

The dual system of the *k*-out-of-*n*:G system is the *k*-out-of-*n*:F system. The maximal state vectors of the dual system are the minimal state vectors of the original system. It is a difficult problem to make it clear how the minimal state vectors of the dual system are determined from the minimal state vectors of the original system, which is essentially a problem of how to derive the maximal state vectors from the minimal state vectors, which remains to be an open problem. We notice that when the system is a binary state system, the dual system of the n-out-of-n:G system is n-out-of-n:F system and is equivalently 1-out-of-n:G system.

When the systems are s-relevant, then the minimal state vectors of $\varphi_{k,n,G}^{-1}[s,\rightarrow)$ are given by (12), then the maximal state vectors of $\varphi_{k,n,G}^{-1}(\leftarrow,s-1]$ are given by those of the series system φ^D, which is given as

$$MA\left(\varphi^D\left(\leftarrow,s-1\right]\right)=\left\{\left((s-1)_i,N^{D\backslash\{i\}}\right)\mid i\in D\right\}$$

and then we have

$$MA\left(\varphi_{k,n,G}\left(\leftarrow,s-1\right]\right)=\left\{\left((s-1)^D,N^{C\backslash D}\right)\mid D\subseteq C, \#D=n-k+1\right\}.$$

5.3 Reliability of a multi-state system

When a probability P is given on the product set Ω_C, which denotes a joint stochastic performance of the components, the image probability $\varphi \circ P$ on S indicates the stochastic performance of the system and it is an important issue to evaluate the probability $\varphi \circ P[s,\rightarrow)$ for $s \in S$. Some evaluation methods have been proposed by many researchers. The key points of the methods are the following relations.

$$s \in S, \quad \varphi \circ P[s,\rightarrow) = P\left(\bigcup_{a \in MI(\varphi^{-1}[s,\rightarrow))} [a,\rightarrow)\right),$$

$$\varphi \circ P(\leftarrow,s] = P\left(\bigcup_{b \in MA(\varphi^{-1}(\leftarrow,s])} (\leftarrow,b]\right),$$

from which we have an exact method by inclusion-exclusion principle, Boolean method, stochastic bounds and approximation methods. The reliability function h_φ of the system φ is easily defined as $h_\varphi = \varphi \circ P$.

5.4 Barlow and Wu system

5.4.1 Definition of Barlow and Wu system

For a binary state system (Ω_C,S,φ), we have the relations (2) and (3), which may be rewritten as

$$x \in \Omega_C, \varphi(x) = \max_{a \in MI\left(\varphi^{-1}(1)\right)} \varphi^{C_1(a)}\left(x^{C_1(a)}\right)$$

$$= \min_{b \in MA\left(\varphi^{-1}(0)\right)} \varphi^{C_0(b)}\left(x^{C_0(b)}\right),$$

where

$$\varphi^{C_1(a)}\left(x^{C_1(a)}\right) = \min_{i \in C_1(a)} x_i,$$

$$\varphi^{C_0(b)}\left(x^{C_0(b)}\right) = \max_{i \in C_0(b)} x_i$$

are respectively the binary series and parallel systems. The definition of k-out-of-n systems are given by referring to these formulae. We also may define a class of multi-state systems by extending these formulae to the multi-state case by using the notions of multi-state series and parallel systems given by Definition 5.1.

Definition 5.3. (Definition of Barlow and Wu system) A system $\left(\varphi_C, S, \varphi\right)$ is called a Barlow and Wu system when there exists a family of sub-sets of C, $\mathcal{D} = \{D_1, ..., D_p\}$, and series systems $\left(\Omega_{D_i}, S, \varphi^{D_i}\right)(D_i \in \mathcal{D})$, such that

(1) $D_i \setminus D_j \neq \phi \, (i \neq j), \cup_{i=1}^{p} D_i = C,$

(2) $x \in \Omega_C$

$$\varphi(x) = \max_{1 \leq i \leq p} \varphi^{D_i}\left(x^{D_i}\right).$$

$\mathcal{D} = \{D_i\}_{i=1}^{p}$ corresponds to the minimal path sets of a binary state system and then uniquely determines a binary state system which is denoted by $\varphi_\mathcal{D}$.

A multi-state system $\left(\Omega_C, S, \varphi\right)$ may be generally decomposed into a family of series systems (Ohi 2016a). Definition 5.3 means that a system is called a Barlow and Wu system when the decomposition is of a particular type.

Remembering that the systems in this chapter are assumed to be coherent and then relevant, when the systems in Definition 5.3 are s-relevant, from Theorem 5.2, we have $\varphi^{D_i}(x) = \min_{j \in D_i} x_j$. Hence, the structure function φ of the definition is written as

$$x \in \Omega_C, \varphi(x) = \max_{D_i \in \mathcal{D}} \min_{j \in D_i} x_j,$$

which is the original formula proposed by Barlow and Wu (1978) in the early period of the study of multi-state systems, called an original

Barlow and Wu system in this article. The minimal state vectors of $\varphi^{-1}(s)$ are given as

$$MI\left(\varphi^{-1}(s)\right)=\left\{\left(s^D,0^{C\backslash D}\right)\mid D\in\mathcal{D}\right\},s\in S,$$

since in this case the state spaces are $\Omega_i = S = \{0, 1,..., N\}$ and for each series system $\varphi^{D_i},(\varphi^{D_i})^{-1}(s)$ has the minimum state vector s^{D_i}.

Setting $\mathcal{D}=\{D\subseteq C\mid\#D=k\}$, Definition 5.3 becomes the k-out-of-n:G system of Definition 5.2.

Original Barlow and Wu systems may be used for modeling systems composed of the same components and having the same physical values, such as pressure, temperature, flow rate and so on.

5.4.2 Stochastic performance of an original Barlow and Wu system composed of i.i.d. components

In the sections 5.4.2, 5.4.3 and 5.4.4, we examine an original Barlow and Wu system. When the components are stochastically independent and identical, i.e., the probability P on Ω_C is the product probability $P=\Pi_{i=1}^n P_i$ and $P_i = P(i \in C)$, where P_i is the probability on Ω_i, and we have

$$P(\varphi\geq s)=\sum_{k=1}^n A_k\left(\varphi_D\right)\left(P[s,\rightarrow)\right)^k\left(P(\leftarrow,s-1]\right)^{n-k}. \tag{14}$$

$A_k\left(\varphi_D\right)$ is defined by (6) for the binary state system φ_D. (14) holds, since writing

$$V_k\left(\varphi,s\right)=\left\{D\subseteq C\mid\#D=k,\varphi\left(\left[s^D,\rightarrow\right)\times\left(\leftarrow,(s-1)^{C\backslash D}\right]\right)\subseteq[s,\rightarrow)\right\},$$

we have

$$\{\varphi\geq s\}=\bigcup_{D\in V_k(\varphi,s)}\left[s^D,\rightarrow\right)\times\left(\leftarrow,(s-1)^{C\backslash D}\right].$$

Similarly to the binary case, supposing

$$\alpha_k=\frac{A_k\left(\varphi_D\right)}{\dbinom{n}{k}}-\frac{A_{k-1}\left(\varphi_D\right)}{\dbinom{n}{k-1}},k=1,2,...,n$$

we have the following theorem from (14).

Theorem 5.3. For $s \in S$, we have the following convex combination for the probability, when the components are stochastically identical and independent.

$$P(\varphi \geq s) = \sum_{k=1}^{n} \alpha_k \binom{m}{k} \left(P[s, \rightarrow) \right)^k \left(P(\leftarrow, s-1]) \right)^{n-k}$$

$$= \sum_{k=1}^{n} \alpha_k P\left(\varphi_{k,n,G} \geq s \right). \tag{15}$$

From Theorem 5.3, it is easily understood that when the components are i.i.d., the stochastic law of the system may be expressed by a convex combination of those of the components;

$$P(\varphi = s) = \sum_{k=1}^{n} \alpha_k P\left(\varphi_{k,n,G} = s \right). \tag{16}$$

We notice that the system is assumed to be an original Barlow and Wu system. Using this convex combination, we examine the entropy of an original Barlow and Wu system and the optimal structure of a safety monitoring system.

5.4.3 Entropy of Barlow-Wu system

Supposing a probability P on Ω_C to be given similarly to the previous section 5.4.2, the entropy of the original Barlow and Wu system is given as

$$H_\varphi(P) = \sum_{s \in S} g\left(P(\varphi = s) \right),$$

where $g(x) = -x \log x$ and is concave on $[0,1]$. Inserting (16) and using the concave property of g, we have

$$H_\varphi(P) \geq \sum_{s \in S} \sum_{k=1}^{n} \alpha_k g\left(P\left(\varphi_{k,n,G} = s \right) \right)$$

$$= \sum_{k=1}^{n} \alpha_k \sum_{s \in S} g\left(P\left(\varphi_{,n,G} = s \right) \right)$$

$$= \sum_{k=1}^{n} \alpha_k H_{\varphi_{k,n,G}}(P)$$

$$\geq \min_{1 \leq k \leq n} H_{\varphi_{k,n,G}}(P).$$

Then we have the following theorem:

Theorem 5.4. A system which minimizes the entropy is given by a k-out-of-n:G system.

5.4.4 *Optimal structure of multi-state Barlow and Wu safety monitoring system*

Similarly to the binary state case of Section 4.4.1, we formulate the safety monitoring problem by Barlow and Wu. E is the state space of the environment monitored by a safety monitoring system modelled by an original Barlow and Wu multi-state system (Ω_C, S, φ). A probability U is given on (E, E) and indicates the probability of an event occurring in the environment. A cost function R is defined on $E \times S$, where S is the state space of the system, and $R(e,s)$ $(e \in E, s \in S)$ denotes a cost incurred when the system's output is s for the environmental state e. The output is determined by the structure function φ, based on the information $x \in \Omega_C$ given by the censors. $P(\cdot,\cdot)$ is a transition function from (E, E) to $(\Omega_C, \mathcal{P}(\Omega_C))$. In this section, for each $e \in E$ we assume the probability $P(e, .)$ to be the product of the probabilities $P_i(e, \cdot)$ on Ω_i $(i \in C)$, where P_i is a transition probability of the censor i from (E, E) to $(\Omega_i, \mathcal{P}(\Omega_i))$. Furthermore, we suppose that the censors are identical, i.e. $P_i(\cdot,\cdot) = P(\cdot,\cdot), i = 1, \cdots, n$. Then, for the structure function φ, the expectation $E_\varphi[R]$ is given as, remenbering the convex combination (16),

$$E_\varphi[R] = \int_{e \in E} \sum_{s \in S} R(e,s) P\left(e, \varphi^{-1}(s)\right) U(de)$$

$$= \int_{e \in E} \sum_{s \in S} R(e,s) \sum_{k=1}^{n} \alpha_k P\left(e, \varphi_{k,n,G}^{-1}(s)\right) U(de)$$

$$= \sum_{k=1}^{n} \alpha_k \int_{e \in E} \sum_{s \in S} R(e,s) P\left(e, \varphi_{k,n,G}^{-1}(s)\right) U(de)$$

$$= \sum_{k=1}^{n} \alpha_k E_{\varphi_{k,n,G}}[R]$$

$$\geq \min_{1 \leq k \leq n} E_{\varphi_{k,n,G}}[R] \qquad (17)$$

Hence, we have the following theorem:

Theorem 5.5. (Optimal Barlow and Wu structure under the cost criterion) An original Barlow and Wu system which minimizes the cost is given by a k-out-of-n: G system, when the censors are stochastically independent and identical. The optimal k is given as the one attaining the minimum (17).

For the conditional entropy of the system φ with respect to the environment E, noticing the concavity of g, we have the

$$H[\varphi \mid E] = \int_{e \in E} \sum_{s \in S} g\left(P(e, \varphi^{-1}(s))\right) U(de)$$

$$\geq \int_{e\in E} \sum_{s\in S} \sum_{k=1}^{n} \alpha_k g\left(P\left(e, \varphi_{k,n,G}^{-1}(s)\right)\right) U(de)$$

$$= \sum_{k=1}^{n} \alpha_k \int_{e\in E} \sum_{s\in S} g\left(P\left(e, \varphi_{k,n,G}^{-1}(s)\right)\right) U(de)$$

$$= \sum_{k=1}^{n} \alpha_k H\left[\varphi_{k,n,G} \mid E\right]$$

$$\geq \min_{1\leq k\leq n} H\left[\varphi_{k,n,G} \mid E\right] \tag{18}$$

Theorem 5.6. (Optimal Barlow and Wu structure under the entropy criterion) An original Barlow and Wu system which minimizes the conditional entropy is given by a k-out-of-n:G system, when the censors are stochastically independent and identical. The optimal k is given as the one attaining the minimum (18).

6. Concluding remarks

In this chapter we show structural and probabilistic examinations of a k-out-of-n:G system in both of binary and multi-state cases and a solution for a practical problem of how to construct an optimal structure of a safety monitoring system modelled by an original Barlow and Wu system. The optimal structure of safety monitoring system is shown to be a k-out-of-n:G system under the cost and entropy criteria. These examinations are, however, performed under the assumption that the components are stochastically identical and independent, since the theorem is based on the fact that the stochastic law of the system is given by the convex combination of the stochastic laws of the k-out-of-n:G (k = 1, 2, ..., n) systems under the above assumptions. The discussion for the entropy criterion in the multi-state case is newly given in this article. Independence assumption is easily generalized to the exchangeable case, a kind of dependent but identical case. Giving an optimal structure when the components are not necessarily identical remains to be solved in future.

REFERENCES

Ansell, J. and A. Bendel. 1982. On the optimality of k-out-of-n:G systems. IEEE Transaction of Reliability R-31: 206–210.

Barlow, R.E. and F. Proschan. 1975. Statistical Theory of Reliability and Life Testing. Holt, Rinehart and Winston, New York.

Barlow, R.E. and A.S. Wu. 1978. Coherent systems with multistate components. Mathematics of Operations Research 3: 275–281.

Birnbaum, Z.W., J.D. Esary and S.C. Saunder. 1961. Multi-component systems and structures and their reliability. Technometrics 3: 55–77.

Birnbaum, Z.W. and J.D. Esary. 1965. Modules of coherent binary systems. SIAM J. Appl. Math. 13: 444–462.

Bodin, L.D. 1970. Approximations to system reliability using a modular decomposition. Technometrics 12: 335–344.

El-Neweihi, E., F. Proschan and J. Sethurman. 1978. Multistate coherent systems. J. Appl. Probability 15: 675–688.

Esary, J.D. and F. Proschan. 1963. Coherent structures of non-identical components. Technometrics 5: 191–209.

Griffith, E.S. 1980. Multistate reliability models. Journal of Applied Probability 17: 735–744.

Hagihara, H., F. Ohi and T. Nishida. 1986a. An optimal structure of safety monitoring systems. Math. Japonica 31: 389–397.

Hagihara, H., M. Sakurai, F. Ohi and T. Nishida. 1986b. Application of Barlow-Wu systems to safety monitoring systems. Technology Reports of The Osala University 36: 245–248.

Huang, J., M.J. Zuo and Z. Fang. 2003. Multi-state consecutive-k-out-of-n systems. IIE Transactions 35: 527–534.

Inoue, K., T. Koda, H. Kumamoto and I. Takami. 1982. Optimal structure of sensor system with two failure modes. IEEE Transactions on Reliability R-31: 119–120.

Kohda, T., K. Inoue, H. Kumamoto and I. Takami. 1981. Optimal logical structure of safety monitoring systems with two failure modes. Transaction of the Society of Instrument and Control Engineers of Japan 17: 36–41. (In Japanese)

Levitin, G. 2013. Multi-state vector-k-out-of-n systems. IEEE Transactions on Reliability 62(3): 648–657.

Lisnianski, A. and G. Levitin. 2003. Multi-State Systems Reliability. Assessment, Optimization and Applications. World Scientific Publishing Co. Pte. Ltd.

Lisnianski, A., I. Frenkel and Y. Ding. 2010. Multi-state System Reliability Analysis and Optimization for Engineers and Industrial Managers. Springer.

Mine, H. 1959. Reliability of physical system. IRE Special Supplement, CT-6: 138–151.

Nakajima, K. and K. Yamato. 1987. On optimal redundancy of multivalue-output systems. IEEE Transactions on Reliability R-36 (2): 216–221.

Natvig, B. 1982. Two suggestions of how to define a multistate coherent system. Adv. Appl. Prob. 14: 434–455.

Natvig, B. 2011. Multistate Systems Reliability Theory with Applications. Wiley.

Ohi, F. and T. Nishida. 1983. Generalized multistate coherent systems. J. Japan Statist. Soc. 13: 165–181.

Ohi, F. and T. Nishida. 1984a. On multistate coherent systems. IEEE Transactions on Reliability R-33: 284–288.

Ohi, F. and T. Nishida. 1984b. Multistate systems in reliability theory, stochastic models in reliability theory. Lecture Notes in Economics and Mathematical Systems 235: 12–22, Springer-Verlag.

Ohi, F., S. Shinmori and T. Nishida. 1989. A definition of associated probability measures on partially ordered sets. Math. Japonica 34: 403–408.

Ohi, F. and T. Suzuki. 2000. Entropy and safety monitoring systems. Japan Journal of Industrial and Applied Mathematics 17: 59–71.

Ohi, F. 2010. Multistate coherent systems. pp. 3–34. *In*: Nakamura, S. and Nakagawa, T. (eds.). Stochastic Reliability Modeling, Optimization and Applications. World Scientific Publishing Co. Pte. Ltd.

Ohi, F. 2012. Multi-State Coherent Systems and Modules – Basic Properties. pp. 374–381. *In*: Yamamoto, H., Qian, C., Cui, L. and Dohi, T. (eds.). The 5th Asia-Pacific International Symposium on Advanced Reliability and Maintenance Modeling, *In*: Advanced Reliability and Maintenance Modeling V, Basis of Reliability Analysis, Nanjing, China, 1–3 November 2012. McGraw Hill Education, Taiwan.

Ohi, F. 2013. Lattice set theoretic treatment of multi-state coherent systems. Reliability Engineering and System Safety 116: 86–90.

Ohi, F. 2015. Steady-state bounds for multi-state systems' reliability via modular decompositions. Applied Stochastic Models in Business and Industry 31: 307–324.

Ohi, F. 2016a. Decomposition of a multi-state systems by series systems. Journal of the Operations Research Society of Japan 59: 291–311.

Ohi, F. 2016b. Stochastic evaluation methods of a multi-state system via a modular decomposition. Journal of Computational Science 17: 156–169.

Phillips, M.J. 1976. The reliability of two-terminal parallel-series networks subject to two kinds of failure. Microelectronics and Reliability 15: 535–549.

Phillips, M.J. 1980. *k-out-of-n:G* systems are preferable. IEEE Transactions of Reliability R-29: 166–169.

Tanaka, K. 1995. Mathematical models of safety monitoring system considering uncertainty. OYO SURI 5: 2–13. (In Japanese)

Yu, K., I. Koren and Y. Guo. 1994. Generalized multistate monotone coherent systems. IEEE Transactions on Reliability 43: 242–250.

An Analysis of (*k*+1)-out-of-*n*:F Fault Tolerant System with Fixed Warranty Period

Nupur Goyal* and Mangey Ram

Department of Mathematics, Graphic Era (Deemed to be University),
Dehradun, Uttarakhand – 248002, India

1. Introduction

In the modern scenario, the multi-unit repairable system has been used in high-tech industries instead of a single unit system. It is crucial for many industries that these systems achieve their aim. These systems have been designed to safety and reliability requirements with redundancy to be fault coverage, due to lack of time, their affordability and estimated reliability of the system (Rahman 2014). Due to the huge investment on the repairable system, the warranty and its proper maintenance are scrupulously required and also take care of cost benefits (Sharma et al. 2009). In order to attain maximum profit, it is also necessary that the system has a fixed warranty period and that it be maintained or serviced by an expert engineer because these failures cause much inconvenience in addition to their economic impact. System reliability is reliant upon proper utilization and maintenance. It is vital for the customer/user that the system is how much reliable and which type of maintenance, it require. Users have greater apprehension because they have become extravagant due to the sophistication of the systems. A fixed warranty period offered by manufacturers gives the users peace of mind. In addition, after the analysis of the system failures, they can ascertain the required type of

---

*Corresponding author: drmrswami@yahoo.com

maintenance and repair facility (Ebeling 2004, Kadyan 2013, Yusuf and Koki 2013, Ram et al. 2013).

There is extensive literature on the reliability of systems in history, but recently, the reliability of a (k+1)-out-of-*n* system has been the focus of special attention because of its wide range of applications in several systems in various fields, such as engineering systems, nuclear power plants, communication systems, etc. Many researchers (Wang and Loman 2002, Koucky 2003, Amari et al. 2008, Angus, 1988, Yusuf and Hussaini 2012, Li and Zuo 2008, Salehi et al. 2011, Arulmozhi 2003) have investigated the reliability measures of the *k*-out-of-*n* structures with different distributions of time and distinct configurations, either parallel or series. They studied the system using a different probabilistic approach, but they did not consider the warranty time and covered faults technique. A parallel configuration of consecutive *k*-out-of-*n*:F system is discussed by Eryilmaz (2014) with the consideration of random deterioration over the course of a lifetime. Wu et al. (2014) analyzed the reliability indices of the *k*-out-of-*n*:G structure of a repairable system and considered the single repair facility that can go to the single vacation. They explained their study by taking the value of *k* as unity. Cekyay and Ozekici (2015) also studied the coherent system with the series connection of the *k*-out-of-*n* structure of the standby systems and assumed that the component lifetimes are distributed exponentially. They derived the generalized expressions of reliability measures.

After faults detection, covering these faults in the repairable system has a great importance in both the system performance and economic impact. Akhtar (1994) modeled the *k*-out-of-*n*:G systems using Markov chains examined the reliability and availability of the system under the consideration of covered faults perfectly or imperfectly. How to calculate the coverage factor for any system is discussed by the Amer and McCluskey (1987). To explain the concept, they developed a simple fault tolerant system and analyzed its reliability with CARE III. Wang et al. (2012) compared the availability of two systems with warm standby units under the assumption that the coverage factor of the active unit failure is different from that of the standby unit failure.

Keeping the above facts in view, the authors have studied a general (k+1)-out-of-*n*:F-structure of the system which is similar to *k*-out-of-*n*:F structure. The difference is only that for (k+1)-out-of-*n*:F structure, (k+1) failed units resulted in complete system failure, and if k units have failed then the system works, but for *k*-out-of-*n* structure, if *k* units have failed completely then the system fails completely. A (k+1)-out-of-*n* failed system is the most prevalent type of redundancy and ordinarily used in engineering systems. The probability of successful operation of the (k+1)-out-of-*n*:F complex system has a great importance in the determination of performance level and quality of the system (Ram 2013). It is worthwhile to highlight here that the system's functionality fluctuates due to the existence of distinct failures in it. It is supposed that the working of the designed $(k+1)$-out-of-n:F system is affected by four different types of failures, namely minor failure, major failure, common cause failure and equipment failure, and that on account of these failures, some faults exist in the system. When the system or sub-system fails due to a negligible cause at any instant, then the failure is known as a minor failure. It can occur due to several factors, such as functional defects, poor design, operation and maintenance conditions, etc. Major failure, however, is a failure that fails the whole system instantaneously and cannot be negligible, e.g. an earthquake or other external catastrophe, fire damage, etc. In the engineering complex repairable system, there is too great a chance of the existence of a common cause failure. It is the failure when the system fails due to any common problem, such as poor maintenance (Jordan and Marshall 1972, Dhillon 1992, Ram and Singh 2010). To measure the reliability characteristics of the system, it is noteworthy to stringently evaluate all the failures or faults of the system and scrutinize the number of faults that can be covered. The greater number of fault coverage factors enhances the system reliability and reduces the maintenance cost. Coverage factor can be defined as the total number of covered faults to the total number of detecting faults (Dugan and Trivedi 1989). The coverage directly impacts the reliability; it is imperative to characterize the recovered fault proportion in order to accurately determine the coverage factor (the proportion of faults from which the system can automatically recover) of redundant equipment and to calculate the impact of the parameter on the overall system reliability. The authors provide the mathematical model of the designed $(k+1)$ -out-of-n:F system, employing birth-death Markov process, supplementary variable technique (Cox 1955, Oliveira et al. 2005), fault coverage technique (Arnold 1973, Pham 1993) and Laplace transformation. The proposed probabilistic model has the following silent topographies:

- The presented model obtains the expression of the probability of each transition state of the designed system as a function of failure rates.
- The discussed model investigated the probabilistic analysis of the system, i.e. availability, reliability and mean time to failure.
- The cost-benefit analysis is also examined in the developed system with the consideration of warranty.
- The suggested model deliberated a strong mathematical establishment with an intuitive graphical representation.

The rest of the paper is organized as follows: Section 2 gives the materials and method, section 3 comprises the mathematical foundation of the model. Probabilistic analyses and their numerical illustrations are shown in section 4. The conclusion of the overall study is explained in section 5.

2. Materials and method

2.1 Assumptions and nomenclature

Throughout the conversation of the model, notations are described in Table 1 and the following suppositions are taken by the authors

- (i) At the initial stage, all the n units of the system work without any failure.
- (ii) Failure rates are taken to be constant.
- (iii) A single repair facility is available always.
- (iv) Repair rates are distributed generally and have an arbitrary value.
- (v) When the system goes to downstate, a repairman takes action immediately.
- (vi) After repair, the system works like a new one.
- (vii) No damage to the system occurs during repair.
- (viii) The cost of repair within the warranty is borne by the manufacturer and after the completion of warranty of the system, the user pays the service cost.

Table 1: Notations

Notation	Description
t/s	Time scale/Laplace transform variable.
$\bar{P}(s)$	Laplace transformation of $P(t)$.
$P_i(t)$	The probability of good or partially failed state S_i of the system; $i = 0, 1, 2, 3$.
$P_j(\sigma, t)$	The probability density function of the complete failed state S_j of the system at epoch time t and has an elapsed repair time of σ; $j = 4, 5, 6, 7$: $\sigma = x, y, z, q$.
λ_k/λ_{k+1}	The constant failure rate for k units/more than k units of the system.
$\lambda_{mi}/\lambda_{mj}$	Minor/ major failure rate
λ_{cc}	Constant rate of common cause failure
β	Failure rate of the system within warranty
α	Rate of completion of warranty period
$\phi_i(\sigma)$	Repair rate of the complete failed state S_j of the system; $i = 1, 2, 3, 4$; $\sigma = q, z, y, x; j = 7, 6, 5, 4$.
c	Coverage parameter
$P_{up}(t)$	Upstate system probability at time t or availability of the system
$Rl(t)$	The reliability of the system at time t.
$E_p(t)$	Expected profit during the interval $(0, t]$.
K_1/K_2	Revenue/service cost per unit time respectively.
w	Warranty time period

2.2 System description

Apart from the importance of reliable systems, reliability characteristics play a key role in the performance of any engineering system. In order to maintain its importance, the authors have proposed a multi-unit system with $(k+1)$-out-of-n:F redundancy. The modeled system consists of n identical repairable units connected to each other, as shown in Fig. 1, and follows $(k+1)$-out-of-n:Failed policy. $(k+1)$-out-of-n:F structure is equivalent to $(n-k)$-out-of-n:G structure of the system. k-out-of-n:G structure of the system is defined as: At least k units should be working properly for the system to be functional. The reliability measures are evaluated by the analysis of distinct failures that occur in the $(k+1)$-out-of-n:F system. The system is studied with the consideration of four types of failures, such as unit failure, common cause failure, minor failure and major failure. With the existence of any of these failures, the system goes to the degraded or completely failed state. In the degraded state, the system can be operating but it works with reduced efficiency, meaning it is failed partially; the system stops working in the completely failed state. Initially, the designed system is considered as being in good working condition and, after the completion of the warranty period, the designed system still remains in good condition.

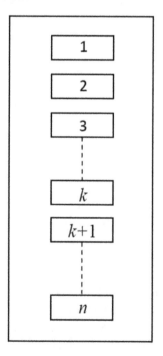

Fig. 1: Configuration diagram

The $(k+1)$-out-of-n:F system goes to completely failed state in the following ways:

(i)   More than k units of the system fail
(ii) Common cause failure occurs in the system
(iii) A major failure occurs in the system
(iv) The system fails within the warranty

The $(k+1)$-out-of-n:F system is considered partially failed in the following ways:

(i) The system fails due to minor failure
(ii) k units of the system have failed

All the operable and inoperable states are shown in Fig. 2 and described in Table 2. A single repairman has been always with the system to repair the failed system. The faults occurring in the system can be detected and proactively repaired before the complete failure of the system, otherwise the component must be repaired or replaced after the failure of the system. Repair within warranty does not affect the user's profit because, within warranty, the repair cost is borne by the manufacturer. The following reliability measures have been analyzed using the supplementary variable technique, Markov process, Laplace transformation and coverage factor.

(i)   Probability of each possible transition state
(ii) Reliability measures namely availability, reliability and mean time to failure
(iii) Expected profit of the system

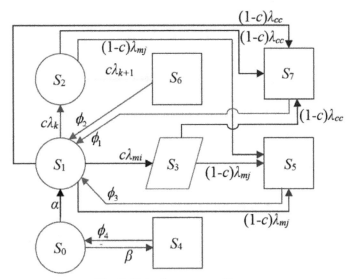

Fig. 2: State changeover diagram

Table 2: State description

State	Description
S_0	State of successful operation.
S_1	Good state of the system after completed the warranty period
S_2	Good working state of the system when k units have failed.
S_3	Degraded state of the system after a minor failure.
S_4	State of the failed system under warranty.
S_5	Due to the major failure, the system is in complete failed state.
S_6	Completely failed state of the system when k+1 or more units have failed.
S_7	Completely failed state of the system due to common cause failure.

3. Mathematical foundation of the model

3.1 Formulation of the model

The following set of differential equations, by the probability of consideration and steadiness of influences, possesses the present mathematical model using Markov process (See Appendix).

$$\left[\frac{\partial}{\partial t} + \alpha + \beta\right]P_0(t) = \int_0^\infty \phi_4(x)P_4(x, t)dx \tag{1}$$

$$\left[\frac{\partial}{\partial t} + c\lambda_k + c\lambda_{mi} + (1-c)\lambda_{mj} + (1-c)\lambda_{cc}\right]P_1(t)$$

$$= \alpha P_0(t) + \int_0^\infty \phi_1(q)P_7(q, t)dq + \int_0^\infty \phi_2(z)P_7(z, t)dz$$

$$+ \int_0^\infty \phi_3(y)P_5(y, t)dy \tag{2}$$

$$\left[\frac{\partial}{\partial t} + c\lambda_{k+1} + (1-c)\lambda_{mj} + (1-c)\lambda_{cc}\right]P_2(t) = c\lambda_k P_1(t) \tag{3}$$

$$\left[\frac{\partial}{\partial t} + (1-c)\lambda_{mj} + (1-c)\lambda_{cc}\right]P_3(t) = c\lambda_{mi}P_1(t) \tag{4}$$

$$\left[\frac{\partial}{\partial t} + \frac{\partial}{\partial x} + \xi\right]P_i(\sigma, t) = 0; \quad i = 4, 5, 6, 7; \quad \xi = \phi(\sigma);$$

$$j = 4, 1, 2, 3; \quad \sigma = x, q, z, y \tag{5}$$

Boundary conditions

$$P_{i+4}(0, t) = \varsigma P_{i+4}(t); \, i = 0, 2; \varsigma = \beta, c\lambda_{k+1} \tag{6}$$

$$P_i(0, t) = (1 - c)\varsigma \sum_{j=1}^{3} P_j(t); \, i = 5, 7; \varsigma = \lambda_{mj}, \lambda_{cc} \tag{7}$$

Initial condition

$$P_i(0) = \begin{cases} 1, & i = 0 \\ 0, & i \geq 1 \end{cases} \tag{8}$$

3.2 Solution of the model

Taking the Laplace transform of Equations (1) to (7), using initial condition and one obtains-

$$[s + \alpha + \beta]\overline{P}_0(s) = 1 + \int_0^\infty \phi_4(x)\overline{P}_4(x, s)dx \tag{9}$$

$$[s + c\lambda_k + c\lambda_{mi} + (1 - c)\lambda_{mj} + (1 - c)\lambda_{cc}]\overline{P}_1(s)$$

$$= \alpha \overline{P}_0(s) + \int_0^\infty \phi_1(q)\overline{P}_7(q, s)dq + \int_0^\infty \phi_2(z)\overline{P}_7(z, s)dz$$

$$+ \int_0^\infty \phi_3(y)\overline{P}_5(y, s)dy \tag{10}$$

$$[s + c\lambda_{k+1} + (1 - c)\lambda_{mj} + (1 - c)\lambda_{cc}]\overline{P}_2(s) = c\lambda_k \overline{P}_1(s) \tag{11}$$

$$[s + (1 - c)\lambda_{mj} + (1 - c)\lambda_{cc}]\overline{P}_3(s) = c\lambda_{mi} \overline{P}_1(s) \tag{12}$$

$$\left[s + \frac{\partial}{\partial x} + \xi\right]\overline{P}_i(\sigma, s) = 0; \quad i = 4, 5, 6, 7; \quad \xi = \phi(\sigma);$$

$$j = 4, 1, 2, 3; \quad \sigma = x, q, z, y \tag{13}$$

$$\overline{P}_{i+4}(0, s) = \varsigma \overline{P}_{i+4}(s); \quad i = 0, 2; \quad \varsigma = \beta, c\lambda_{k+1} \tag{14}$$

$$\overline{P}_i(0, s) = (1 - c)\varsigma \sum_{j=1}^{3} \overline{P}_j(s); \quad i = 5, 7; \quad \varsigma = \lambda_{mj}, \lambda_{cc} \tag{15}$$

The authors have calculated the Laplace transformation of each possible transition's probability of the system through Equation (9) to (15), as

$$\overline{P}_0(s) = \frac{1}{A_1(s)} \tag{16}$$

$$\bar{P}_1(s) = \frac{\alpha}{A_4(s)} \bar{P}_0(s) \tag{17}$$

$$\bar{P}_i(s) = \frac{\alpha A_i(s)}{A_4(s)} \bar{P}_0(s); \quad i = 2,3 \tag{18}$$

$$\bar{P}_{i+4}(s) = \left(\frac{1 - \bar{S}_{\phi_j}(s)}{s}\right) \xi \bar{P}_0(s); \qquad \xi = \beta,\, c\lambda_{k+1} \frac{\alpha A_2(s)}{A_4(s)};$$
$$i = 0,2; \qquad j = 4,2 \tag{19}$$

$$\bar{P}_i(s) = \frac{\alpha}{A_4(s)} \left(\frac{1 - \bar{S}_{\phi_j}(s)}{s}\right)(1-c)(1 + A_2(s) + A_3(s))\xi \bar{P}_0(s);$$
$$\xi = \lambda_{mj},\, \lambda_{cc}; \qquad i = 5,7; \qquad j = 3,1 \tag{20}$$

where,

$$A_1(s) = s + \alpha + \beta(1 - \bar{S}_{\phi_4}(s)),\, A_2(s) = \frac{c\lambda_k}{s + c\lambda_{k+1} + (1-c)(\lambda_{mj} + \lambda_{cc})},$$

$$A_3(s) = \frac{c\lambda_{mi}}{s + (1-c)(\lambda_{mj} + \lambda_{cc})}$$

$$A_4(s) = s + c(\lambda_k + \lambda_{mi}) + (1-c)(\lambda_{mj} + \lambda_{cc}) - (1-c)\left(1 + A_2(s) + A_3(s)\right)$$
$$(\lambda_{mj}\bar{S}_{\phi_3}(s) + \lambda_{cc}\bar{S}_{\phi_1}(s)) - c\lambda_{k+1}\bar{S}_{\phi_2}(s)A_2(s)$$

3.3 Upstate and downstate system probability

With the help of all transitions, shown in transition diagram and Equations (16) to (20), one obtains the expression of the Laplace transformation regarding the the probability of upstate and downstate systems.

$$\bar{P}_{up}(s) = \sum_{i=0}^{3} \bar{P}_i(s)$$
$$= \left[1 + \frac{\alpha}{A_4(s)}(1 + A_2(s) + A_3(s))\right]\bar{P}_0(s) \tag{21}$$

$$\bar{P}_{down}(s) = \sum_{i=4}^{7} \bar{P}_i(s)$$
$$= \left[\frac{\alpha}{A_4(s)}\left\{(1-c)(1 + A_2(s) + A_3(s))\sum \xi\left(\frac{1 - \bar{S}_{\phi_j}(s)}{s}\right) + \left(\frac{1 - \bar{S}_{\phi_2}(s)}{s}\right)c\lambda_{k+1}A_2(s)\right\}\right.$$
$$\left. + \left(\frac{1 - \bar{S}_{\phi_4}(s)}{s}\right)\beta\right]\bar{P}_0(s); \quad \varsigma = \lambda_{mj},\, \lambda_{cc};\, j = 3,1 \tag{22}$$

4. Particular cases and numerical computations

4.1 Availability analysis

It can be described by the probability of the performance of the required function at any instant or during a given time interval when the system is operated or installed under a prescribed discipline. It depends on the system organization as well as on the component's availability. Availability of the repairable system is the function of both its failure rates and repair or replacement rates (Avizienis et al. 2004, Samrout et al. 2005). Setting the failure and repair rates as $\alpha = 0.03$, $\beta = 0.01$, $\lambda_k = 0.05$, $\lambda_{k+1} = 0.06$, $\lambda_{cc} = 0.01$, $\lambda_{mi} = 0.02$, $\lambda_{mj} = 0.03$, $\phi_1(q) = 0.2$, $\phi_2(z) = 0.3$, $\phi_3(y) = 0.4$, $\phi_4(x) = 0.1$ (Kadyan 2013; Singh et al. 2013) in Equation (21) and by taking the inverse Laplace transformation, one can obtain the availability of the considered $(k+1)$-out-of-n:F system, as shown in Fig. 3, after varying time unit t from 0 to 15 and setting the coverage parameter $c = 0.1, 0.3, 0.5, 0.7, 0.9$, respectively.

From the graphs shown in Fig. 3, one can observe that the availability of the system is initially the same at any value of coverage factor but, after a short period of time, decreases in similar a manner at a different value. Availability of the considered system attains a higher value as the coverage factor increases.

4.1.1 *Limiting expression of availability*

By using Abel's Lemma in the Laplace transformation, that is to say:

$$\underset{s \to 0}{Lim}\left[s\overline{P}(s)\right] = \underset{t \to \infty}{Lim}\left[P(t)\right] = P$$

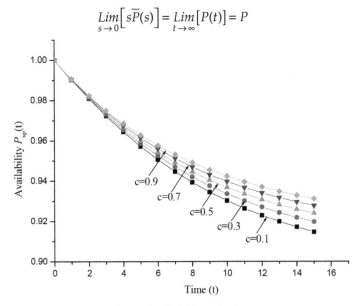

Fig. 3: Availability v/s time

Where, $F(t)$ is a function of time t and $\overline{P}(s)$ is the Laplace transform of $F(t)$, provided that the limit on right hand exists, the limiting expression of availability of designed is obtained as follows:

$$P = \left[1 + \frac{\alpha}{A_4}(1 + A_2 + A_3) \right] \frac{1}{A_1}$$

where,

$$A_1 = \alpha + \beta(1 - \int x\phi_4(x)dx), \quad A_2 = \frac{c\lambda_k}{c\lambda_{k+1} + (1-c)(\lambda_{mj} + \lambda_{cc})},$$

$$A_3 = \frac{c\lambda_{mi}}{(1-c)(\lambda_{mj} + \lambda_{cc})}$$

$$A_4 = c(\lambda_k + \lambda_{mi}) + (1-c)(\lambda_{mj} + \lambda_{cc}) - (1-c)(1 + A_2 + A_3)$$

$$(\lambda_{mj} \int y\phi_3(y)dy + \lambda_{cc} \int q\phi_1(q)dq) - c\lambda_{k+1} \int z\phi_2(z)dzA_2$$

4.2 Reliability analysis

It is defined as the probability that a system will perform its intended function during a specified period of time under stated conditions. It is always a function of time. It also depends on environmental conditions, which may or may not vary with time (Dhillon and Singh 1981, Meeker and Escobar 2004, Goyal et al. 2016a). Repair facility was not taken into account for the reliability evaluation. Setting all repair rates to zero and assigning the value of failure rates as $\alpha = 0.03$, $\beta = 0.01$, $\lambda_k = 0.05$, $\lambda_{k+1} = 0.06$, $\lambda_{cc} = 0.01$, $\lambda_{mi} = 0.02$, $\lambda_{mj} = 0.03$ (Kadyan 2013, Singh et al. 2013) in Equation (21), later on taking the inverse Laplace transformation. One can get the reliability expression in terms of time and coverage factor as

$$Rl(t) = \frac{1}{c}\left[\begin{array}{l} (-7.5)e^{(-0.02(c+2)t)} + 0.2142857143e^{(0.04(c-1)t)} \\ + (c+3)e^{(-0.04\,t)} + 4.285714286e^{(-0.01(3c+4)t)} \end{array} \right] \tag{23}$$

Varying the time unit t from 0 to 15 and setting the coverage parameter $c = 0.01, 0.03, 0.05, 0.07, 0.09$ respectively, one can determine the reliability of the considered system, as revealed in Fig. 4.

The critical examination of graphs reveals that, at the starting time period, the system is equally reliable at distinct values of coverage factor and, later on, the reliability of the system is decreased in a similar discipline at distinct values of coverage. With the passage of time, the reliability of the considered system decreases and the probability of failure of the system increases. The system is more reliable at higher coverage factor as compared to low.

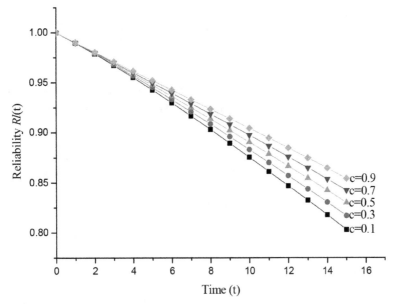

Fig. 4: Reliability v/s time

4.3 MTTF analysis

MTTF of considered $(k+1)$-out-of-n:F repairable system is the predicted elapsed time between characteristic failures of the system or sub-system during operation. MTTF is the average time between failures of a system or sub-system (Ram and Manglik, 2014; Goyal et al. 2016b). Consider that the repair facility is not available, putting each repair rate as zero and taking the limits as s tends to zero in Equation (21), the expected mean time to failure of the considered $(k+1)$-out-of-n:F repairable system is

$$MTTF = \frac{1}{\alpha+\beta}\left[1 + \frac{\alpha\left(1 + \frac{c\lambda_k}{c\lambda_{k+1}+(1-c)(\lambda_{cc}+\lambda_{mj})} + \frac{c\lambda_{mi}}{(1-c)(\lambda_{cc}+\lambda_{mj})}\right)}{c(\lambda_k+\lambda_{mi})+(1-c)(\lambda_{cc}+\lambda_{mj})}\right] \quad (24)$$

By putting the coverage parameter $c = 0.1, 0.3, 0.5, 0.7, 0.9$, respectively, in Equation (24), then the MTTF of the considered repairable system corresponding to different possibilities can be predicted by Equation (24) as follows:

(i) Taking $\beta = 0.01$, $\lambda_k = 0.05$, $\lambda_{k+1} = 0.06$, $\lambda_{cc} = 0.01$, $\lambda_{mi} = 0.02$, $\lambda_{mj} = 0.03$ and varying $\alpha = 0.01, 0.02, 0.03, 0.04, 0.05, 0.06, 0.07, 0.08$ and 0.09 respectively, one can estimate the MTTF of the system, as represented in Fig. 5.

(ii) Setting $\alpha = 0.03$, $\lambda_k = 0.05$, $\lambda_{k+1} = 0.06$, $\lambda_{cc} = 0.01$, $\lambda_{mi} = 0.02$, $\lambda_{mj} = 0.03$, and fluctuating $\beta = 0.01, 0.02, 0.03, 0.04, 0.05, 0.06, 0.07, 0.08$ and 0.09, respectively, one can evaluate the MTTF of the system, as shown in Fig. 6.

(iii) Fixing $\alpha = 0.03$, $\beta = 0.01$, $\lambda_{k+1} = 0.06$, $\lambda_{cc} = 0.01$, $\lambda_{mi} = 0.02$, $\lambda_{mj} = 0.03$, and varying $\lambda_k = 0.01, 0.02, 0.03, 0.04, 0.05, 0.06, 0.07, 0.08$ and 0.09, respectively, one can calculate the MTTF of the system, as revealed in Fig. 7.

(iv) Substituting $\alpha = 0.03$, $\beta = 0.01$, $\lambda_k = 0.05$, $\lambda_{cc} = 0.01$, $\lambda_{mi} = 0.02$, $\lambda_{mj} = 0.03$, and varying $\lambda_{k+1} = 0.01, 0.02, 0.03, 0.04, 0.05, 0.06, 0.07, 0.08$ and 0.09, respectively, one can estimate the MTTF of the system, as shown in Fig. 8.

(v) Putting $\alpha = 0.03$, $\beta = 0.01$, $\lambda_k = 0.05$, $\lambda_{k+1} = 0.06$, $\lambda_{mi} = 0.02$, $\lambda_{mj} = 0.03$, and varying $\lambda_{cc} = 0.01, 0.02, 0.03, 0.04, 0.05, 0.06, 0.07, 0.08$ and 0.09, respectively, one can estimate the MTTF of the system, as revealed in Fig. 9.

(vi) Taking $\alpha = 0.03$, $\beta = 0.01$, $\lambda_k = 0.05$, $\lambda_{k+1} = 0.06$, $\lambda_{cc} = 0.01$, $\lambda_{mj} = 0.03$, varying $\lambda_{mi} = 0.01, 0.02, 0.03, 0.04, 0.05, 0.06, 0.07, 0.08$ and 0.09, respectively, one can estimate the MTTF of the system, as publicised in Fig. 10.

(vii) Substituting $\alpha = 0.03$, $\beta = 0.01$, $\lambda_k = 0.05$, $\lambda_{k+1} = 0.06$, $\lambda_{cc} = 0.01$, $\lambda_{mi} = 0.02$, varying $\lambda_{mj} = 0.01, 0.02, 0.03, 0.04, 0.05, 0.06, 0.07, 0.08$ and 0.09, respectively, one can calculate the MTTF of the system, as revealed in Fig. 11.

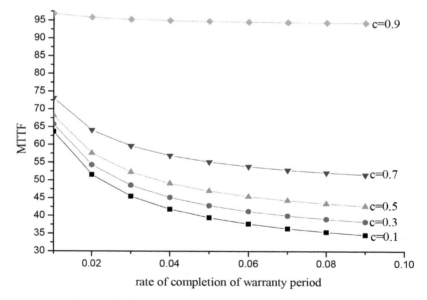

Fig. 5: MTTF v/s rate of completion of warranty period

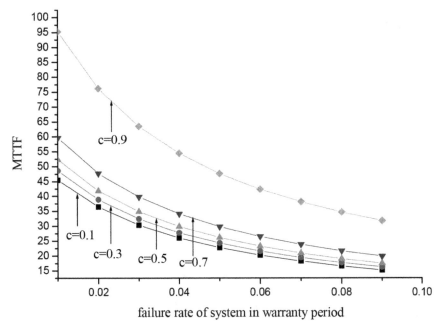

Fig. 6: MTTF v/s failure rate under warranty

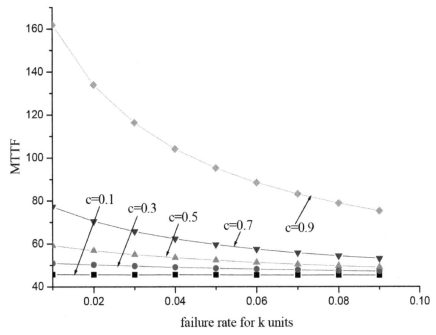

Fig. 7: MTTF v/s failure rate of k units

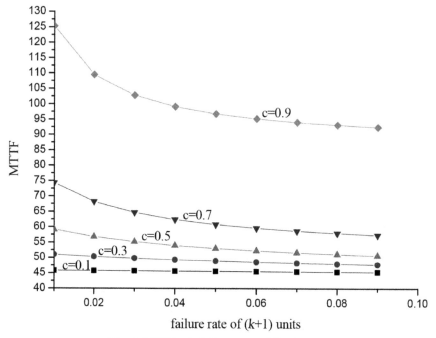

Fig. 8: MTTF v/s failure rate of (k+1) units

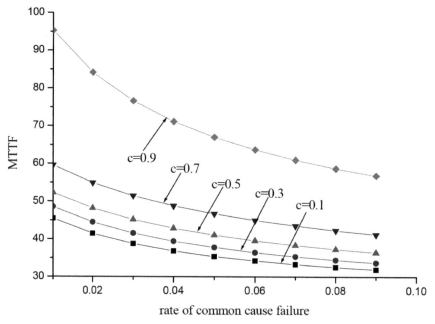

Fig. 9: MTTF v/s rate of common cause failure

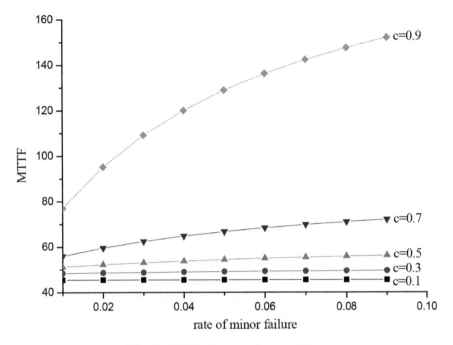

Fig. 10: MTTF v/s rate of minor failure

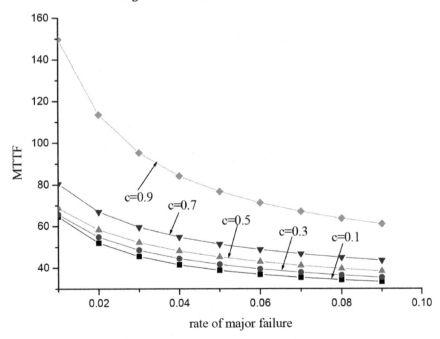

Fig. 11: MTTF v/s rate of major failure

MTTF of the complex system decreases as the rate of completion of warranty period increases. MTTF of the considered complex system decreases in a different manner with respect to the variation in the failure rate of the system under warranty and after warranty, except for minor failure. When the system has failed due to minor failure, MTTF of the system increases with the increment in its failure rate. MTTF of the considered complex system increases by covering more faults. At higher coverage factor, MTTF of the system has a higher value in each situation. Due to minor failure, at $c = 0.3$ and $c = 0.1$, MTTF of the system is approximately the same with the variation in the rate of the minor failure.

4.4 Expected profit

Expected profit of the user for any system depends on upon the warranty period, service cost and failure of its components. Consider that $(0, w]$ is the warranty period of the designed system; since the repair facility is always available, after the completion of the warranty period, the repairman remains busy during the time period $(w, t]$ (Kadyan 2013). Let K_1 be the revenue and K_2 be the service cost per unit time, then the expected profit $E_p(t)$ during the interval $(0, t]$ is given by

$$E_P(t) = K_1 \int_0^t P_{up}(t)dt - (t - w)K_2 \qquad (25)$$

Taking the inverse Laplace transform of Equation (21) and using it in (25), for the distinct value of component's failure and repair rates as $\alpha = 0.03$, $\beta = 0.01$, $\lambda_k = 0.05$, $\lambda_{k+1} = 0.06$, $\lambda_{cc} = 0.01$, $\lambda_{mi} = 0.02$, $\lambda_{mj} = 0.03$, $\phi_1(q) = 0.2$, $\phi_2(z) = 0.3$, $\phi_3(y) = 0.4$, $\phi_4(x) = 0.1$ (Kadyan 2013; Singh et al. 2013) and setting the coverage factor $c = 0.1, 0.3, 0.5, 0.7, 0.9$, respectively, the expected profit of the considered $(k+1)$-out-of-n:F repairable system can be obtained as follows:

$$E_p(t)_{c=0.1} = K_1 \left(0.0475788033 e^{(-0.048114277721\,t)} - 0.0135798087 e^{(-0.03699091107\,t)} \right.$$

$$+ 0.8972415882\,t - 0.0001125045 e^{(-0.2998986564\,t)}$$

$$+ 0.0123652610 e^{(-0.4280910613\,t)} + 0.0302261894 e^{(-0.2079050935\,t)}$$

$$- 1.069230033 \left[\cosh(0.1135889894\,t) - \sinh(0.1135889894\,t) \right]$$

$$+ 0.7896716636 + 0.2030804288 \left[\cosh(0.0264110106\,t) \right.$$

$$\left. + \sinh\,(0.0264110106\,t) \right] \Big) - (t - w)K_2 \qquad (26a)$$

$$E_p(t)_{c=0.3} = K_1 \left(0.0098456199 e^{(-0.4215458772\,t)} - 0.0011957072 e^{(-0.2989782328\,t)} \right.$$

$$+ 0.9089243189\,t + 0.0264382424 e^{(-0.206434508\,t)}$$

$$+ 0.1325811094 e^{(-0.0651220084\,t)} - 0.4144131209 e^{(-0.0309193737\,t)}$$

$$-1.049851306\left[\cosh\left(0.1135889894\,t\right)-\sinh(0.1135889894\,t)\right]$$

$$+0.4217839677+0.8748111943\left[\cosh\left(0.0264110106\,t\right)\right.$$

$$+\sinh\left(0.0264110106\,t\right)\right]\right)-(t-w)K_2 \tag{26b}$$

$$E_p(t)_{c=0.5}=K_1\left(0.0224120143e^{(-0.2049407283\,t)}-0.0040713043e^{(-0.296778317\,t)}\right.$$

$$+0.9195402567\,t+0.1324613599e^{(-0.0834483215\,t)}$$

$$+4.985148177e^{(-0.02477927057\,t)}+0.0070870692e^{(-0.4150533626\,t)}$$

$$-1.003841812\left[\cosh(0.1135889894\,t)-\sinh(0.1135889894\,t)\right]$$

$$-0.074865871-4.064329633\left[\cosh(0.0264110106\,t)\right.$$

$$-\sinh(0.0264110106\,t)\right]\right)-(t-w)K_2 \tag{26c}$$

$$E_p(t)_{c=0.7}=K_1\left(0.0041875605e^{(-0.408741005\,t)}-0.0102731606e^{(-0.2927115714\,t)}\right.$$

$$+0.9348094748\,t+0.0175314999e^{(-0.2033381408\,t)}$$

$$-0.09786740754e^{(-0.1036356475\,t)}+3.636634409e^{(-0.0185736354\,t)}$$

$$-0.7523413297\left[\cosh\left(0.1135889894t\right)-\sinh\left(0.1135889894t\right)\right]$$

$$-1.305856513-1.492015058\left[\cosh(0.0264110106\,t)\right.$$

$$-\sinh(0.0264110106\,t)\right]\right)-(t-w)K_2 \tag{26d}$$

$$E_p(t)_{c=0.7}=K_1\left(0.0013306148e^{(-0.4027822823\,t)}-0.0235361117e^{(-0.28576179\,t)}\right.$$

$$+0.9664147714\,t+0.0092495344e^{(-0.201376927\,t)}$$

$$+0.5871284893e^{(-0.1267735875\,t)}+7.628657865e^{(-0.01230541313\,t)}$$

$$-1.418070909\left[\cosh(0.1135889894\,t)-\sinh(0.1135889894\,t)\right]$$

$$-5.403236759-1.381522724\left[\cosh(0.0264110106\,t)\right.$$

$$-\sinh(0.0264110106\,t)\right]\right)-(t-w)K_2 \tag{26e}$$

Let us consider the revenue of the designed system $K_1 = 500$. The warranty period is (0, 3), and the service cost varies as 150, 100 and 50 respectively in Equations (26), one obtains Figs 12, 13, 14, 15 and 16, which represent the graphs of expected profit for the designed system, with respect to time and service cost at distinct values of coverage factor.

The graphs shown in the Figs 12, 13, 14, 15 and 16, epitomized that the profit of the system much depends on the service cost. Expected profit of the system decreases as service cost of the system increases. The intersecting point of the graphs in each figure shows the completion of the warranty period of the system. Under the warranty period, service cost

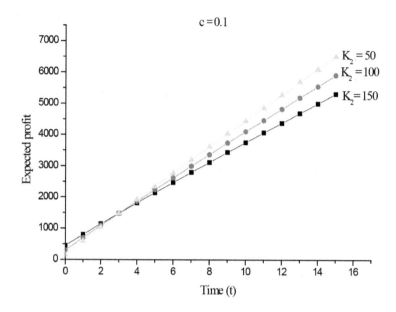

Fig. 12: Expected profit v/s time at $c = 0.1$

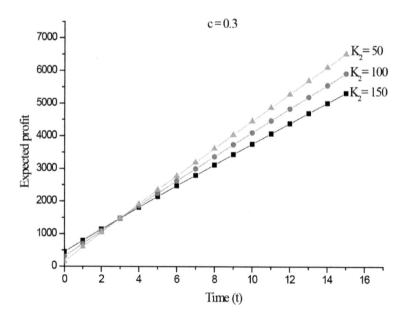

Fig. 13: Expected profit v/s time at $c = 0.3$

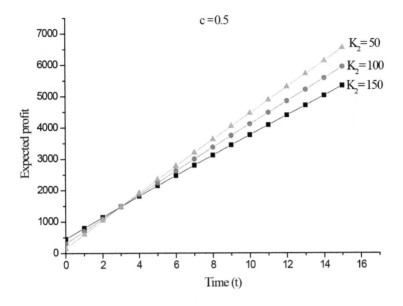

Fig. 14: Expected profit v/s time at $c = 0.5$

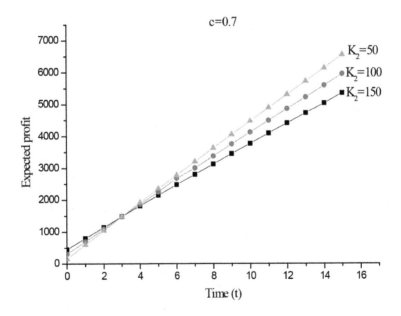

Fig. 15: Expected profit v/s time at $c = 0.7$

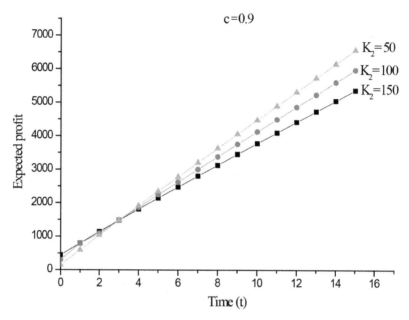

Fig. 16: Expected profit v/s time at $c = 0.9$

of the system affects the expected profit of the manufacturer; it does not affect the user profit. One can also observe that the increment in coverage factor does not greatly affect the user's profit.

5. Conclusion

In a modern scenario, reliability evaluation is an important aspect of designing, safety and fault detection in the complex repairable system. The authors focus on a (k+1)-out-of-n repairable system with failed policy. The performance of the considered is affected by the good or bad working of its components. Thus, the performance characteristics are expressed in terms of component's failure rates. Through the overall analysis, we concluded that the (k+1)-out-of-n repairable system can achieve a higher reliability and perform better by controlling the failure rates of its component and detecting and covering the maximum faults. The evidence composed in this paper is very accommodating for the reliability engineers because to maintain the stability of reliability and other properties, such as mean time to failure, cost, profit, etc., is a significant part of designing a highly reliable, complex system. The nature of this study facilitates the application of reliability theory with a high level of confidence. Further structure will be evaluated with different reliability approaches in the future.

Appendix 1

At any time t, probability that the system is in state S_i at time t and remains there in interval $(t, t + \Delta t)$ or/and if it is in some other state at time t then it should be transited to the state S_j in the interval $(t, t + \Delta t)$, provided transition exists between the states and $\Delta t \to 0$. Accordingly, the equations (1) to (5) are interpreted as given below.

The probability of the system to be in the state S_0 in the interval $(t, t + \Delta t)$ is given by:

$$P_0(t + \Delta t) = (1 - \alpha\Delta t)(1 - \beta\Delta t)P_0(t) + \int_0^\infty P_4(x, t)\phi_4(x)\Delta t dx$$

$$\Rightarrow \lim_{\Delta t \to 0} \frac{P_0(t + \Delta t) - P_0(t)}{\Delta t} + (\alpha + \beta)P_0(t) = \int_0^\infty P_4(x, t)\phi_4(x)\Delta t dx$$

$$\Rightarrow \left[\frac{\partial}{\partial t} + \alpha + \beta\right]P_0(t) = \int_0^\infty P_4(x, t)\phi_4(x)\Delta t dx \qquad \text{(A.1)}$$

For the state S1,

$$P_1(t + \Delta t) = (1 - c\lambda_{mi}\Delta t)((1 - c)\lambda_{cc}\Delta t)(1 - c\lambda_k\Delta t)$$

$$((1 - c)\lambda_{mj}\Delta t)P_1(t) + \int_0^\infty \phi_1(q)\Delta t P_7(q, t)dq$$

$$+ \int_0^\infty \phi_3(y)\Delta t P_5(y, t)dy + \int_0^\infty \phi_2(z)\Delta t P_6(z, t)dz + \alpha\Delta t P_0(t)$$

$$\Rightarrow \lim_{\Delta t \to 0} \frac{P_1(t + \Delta t) - P_1(t)}{\Delta t} + \{c\lambda_{mi} + c\lambda_k + (1 - c)\lambda_{cc}$$

$$+ (1 - c)\lambda_{mj}\}P_1(t) = \int_0^\infty \phi_1(q)\Delta t P_7(q, t)dq$$

$$+ \int_0^\infty \phi_3(y)\Delta t P_5(y, t)dy + \int_0^\infty \phi_2(z)\Delta t P_6(z, t)dz + \alpha\Delta t P_0(t)$$

$$\Rightarrow \left[\frac{\partial}{\partial t} + c\lambda_{mi} + c\lambda_k + (1 - c)\lambda_{cc} + (1 - c)\lambda_{mj}\right]P_1(t) = \int_0^\infty \phi_1(q)\Delta t P_7(q, t)dq$$

$$+ \int_0^\infty \phi_3(y)\Delta t P_5(y, t)dy + \int_0^\infty \phi_2(z)\Delta t P_6(z, t)dz + \alpha\Delta t P_0(t) \qquad \text{(A.2)}$$

For the state S2,

$$P_2(t + \Delta t) = ((1 - c)\lambda_{mj}\Delta t)(1 - c\lambda_{k+1}\Delta t)((1 - c)\lambda_{cc}\Delta t)P_2(t) + c\lambda_k \Delta t P_1(t)$$

$$\Rightarrow \lim_{\Delta t \to 0} \frac{P_2(t + \Delta t) - P_2(t)}{\Delta t} + (c\lambda_{k+1} + (1 - c)\lambda_{mj} + (1 - c)\lambda_{cc})P_2(t) = c\lambda_k P_1(t)$$

$$\Rightarrow \left[\frac{\partial}{\partial t} + c\lambda_{k+1} + (1 - c)\lambda_{mj} + (1 - c)\lambda_{cc} \right] P_2(t) = c\lambda_k P_1(t) \tag{A.3}$$

For the state S3,

$$P_3(t + \Delta t) = ((1 - c)\lambda_{mj}\Delta t)((1 - c)\lambda_{cc}\Delta t)P_2(t) + c\lambda_{mi}\Delta t P_1(t)$$

$$\Rightarrow \lim_{\Delta t \to 0} \frac{P_2(t + \Delta t) - P_2(t)}{\Delta t} + ((1 - c)\lambda_{mj} + (1 - c)\lambda_{cc})P_3(t) = c\lambda_{mi}P_1(t)$$

$$\Rightarrow \left[\frac{\partial}{\partial t} + (1 - c)\lambda_{mj} + (1 - c)\lambda_{cc} \right] P_3(t) = c\lambda_{mi}P_1(t) \tag{A.4}$$

For the state S4,

$$P_4(x + \Delta x, t + \Delta t,) = \{1 - \phi_4(x)\Delta t\}P_4(x, t)$$

$$\Rightarrow \lim_{\substack{\Delta x \to 0 \\ \Delta t \to 0}} \frac{P_4(x + \Delta x, t + \Delta t) - P_4(x, t)}{\Delta t} + \phi_4(x)P_4(x, t) = 0$$

$$\Rightarrow \left[\frac{\partial}{\partial t} + \frac{\partial}{\partial x} + \phi_4(x) \right] P_4(x, t) = 0 \tag{A.5}$$

For the state S5,

$$P_5(y + \Delta y, t + \Delta t,) = \{1 - \phi_3(y)\Delta t\}P_5(y, t)$$

$$\Rightarrow \lim_{\substack{\Delta y \to 0 \\ \Delta t \to 0}} \frac{P_5(y + \Delta y, t + \Delta t) - P_5(y, t)}{\Delta t} + \phi_3(y)P_5(y, t) = 0$$

$$\Rightarrow \left[\frac{\partial}{\partial t} + \frac{\partial}{\partial y} + \phi_3(y) \right] P_5(y, t) = 0 \tag{A.6}$$

For the state S6,

$$P_6(z + \Delta z, t + \Delta t,) = \{1 - \phi_2(z)\Delta t\}P_6(z, t)$$

$$\Rightarrow \lim_{\substack{\Delta z \to 0 \\ \Delta t \to 0}} \frac{P_6(z + \Delta z, t + \Delta t) - P_6(z, t)}{\Delta t} + \phi_2(z)P_6(z, t) = 0$$

$$\Rightarrow \left[\frac{\partial}{\partial t} + \frac{\partial}{\partial z} + \phi_2(z)\right] P_6(z, t) = 0 \tag{A.7}$$

For the state S7,

$$P_7(q + \Delta q, t + \Delta t,) = \{1 - \phi_1(q)\Delta t\} P_7(q, t)$$

$$\Rightarrow \lim_{\substack{\Delta q \to 0 \\ \Delta t \to 0}} \frac{P_7(q + \Delta q, t + \Delta t) - P_7(q, t)}{\Delta t} + \phi_1(q) P_7(q, t) = 0$$

$$\Rightarrow \left[\frac{\partial}{\partial t} + \frac{\partial}{\partial q} + \phi_1(q)\right] P_7(q, t) = 0 \tag{A.8}$$

Appendix 2

Boundary conditions (A.6) to (A.7) within the system are obtained corresponding to transitions between the states where a transition from a state with and without elapsed repair time exists when elapsed repair times x, y, z, q are 0. Hence, from Fig. 2, we have obtained the following boundary conditions:

$$P_4(0, t) = \beta P_0(t) \tag{A.9}$$

$$P_5(0, t) = (1 - c)\lambda_{mj}[P_1(t) + P_2(t) + P_3(t)] \tag{A.10}$$

$$P_6(0, t) = c\lambda_{k+1} P_2(t) \tag{A.11}$$

$$P_7(0, t) = (1 - c)\lambda_{cc}[P_1(t) + P_2(t) + P_3(t)] \tag{A.12}$$

When the system is perfectly good, i.e. in the initial state S_0, then

$$P_0(0) = 1 \text{ and other state probabilities are zero at } t = 0 \tag{A.13}$$

REFERENCES

Akhtar, S. 1994. Reliability of *k*-out-of-*n*:G systems with imperfect fault-coverage. IEEE Transactions on Reliability 43(1): 101–106.

Amari, S.V., M.J. Zuo and G. Dill. 2008. O (kn) Algorithms for analyzing repairable and non-repairable *k*-out-of-*n*:G Systems. pp. 309–320. *In*: Handbook of Performability Engineering. Springer, London.

Amer, H.H. and E.J. McCluskey. 1987. Calculation of coverage parameter. IEEE Transactions on Reliability 2: 194–198.

Angus, J.E. 1988. On computing MTBF for a *k*-out-of-*n*:G repairable system. IEEE Transactions on Reliability 37(3): 312–313.

Arnold, T.F. 1973. The concept of coverage and its effect on the reliability model of a repairable system. IEEE Transactions on Computers 100(3): 251–254.

Arulmozhi, G. 2003. Direct method for reliability computation of k-out-of-n:G systems. Applied Mathematics and Computation 143(2): 421–429.

Avižienis, A., J.C. Laprie, B. Randell and C. Landwehr. 2004. Basic concepts and taxonomy of dependable and secure computing. IEEE Transactions on Dependable and Secure Computing 1(1): 11–33.

Çekyay, B. and S. Özekici. 2015. Reliability, MTTF and steady-state availability analysis of systems with exponential lifetimes. Applied Mathematical Modelling 39(1): 284–296.

Cox, D.R. 1955. The analysis of non-Markovian stochastic processes by the inclusion of supplementary variables. pp. 433–441. In: Mathematical Proceedings of the Cambridge Philosophical Society. Vol. 51, No. 03. Cambridge University Press.

Dhillon, B.S. 1992. Failure modes and effects analysis—bibliography. Microelectronics Reliability 32(5): 719–731.

Dhillon, B.S. and C. Singh. 1981. Engineering Reliability: New Techniques and Applications. Wiley, New York.

Dugan, J.B. and K.S. Trivedi. 1989. Coverage modelling for dependability analysis of fault-tolerant systems. IEEE Transactions on Computers 38(6): 775–787.

Ebeling, C.E. 2004. An Introduction to Reliability and Maintainability Engineering. Tata McGraw-Hill Education, New York.

Eryilmaz, S. 2014. Parallel and consecutive k-out-of-n:F systems under stochastic deterioration. Applied Mathematics and Computation 227: 19–26.

Goyal, N., A.K. Dua and M. Ram. 2016. Autonomous robot performance evaluation with rework. International Journal of Reliability and Safety 10(3): 265–284.

Goyal, N., M. Ram and A. Kaushik. 2016. Performability of solar thermal power plant under reliability characteristics. International Journal of System Assurance Engineering and Management 8(2): 1–9.

Jordan, W.E. and G.C. Marshall. 1972. Failure modes, effects and criticality analyses. pp. 30–37. In: Proceedings of the Annual Reliability and Maintainability Symposium.

Kadyan, M.S. 2013. Cost-benefit analysis of a single-unit system with warranty for repair. Applied Mathematics and Computation 223: 346–353.

Koucký, M. 2003. Exact reliability formula and bounds for general k-out-of-n systems. Reliability Engineering and System Safety 82(2): 229–231.

Li, W. and M.J. Zuo. 2008. Reliability evaluation of multi-state weighted k-out-of-n systems. Reliability Engineering and System Safety 93(1): 160–167.

Meeker, W.Q. and L.A. Escobar. 2003. Reliability: Tthe other dimension of quality. Quality Technology and Qualitative Management 1(1): 1.

Oliveira, E.A., A.C.M. Alvim and P.F. e Melo. 2005. Unavailability analysis of safety systems under aging by supplementary variables with imperfect repair. Annals of Nuclear Energy 32(2): 241–252.

Pham, H. 1993. Optimal cost-effective design of triple-modular-redundancy-with spares systems. IEEE Transactions on Reliability 42(3): 369–374.

Rahman, A. 2014. Maintenance Contract Model for Complex Asset/Equipment. International Journal of Reliability, Quality and Safety Engineering 21(01): 1450002.

Ram, M. 2013. On system reliability approaches: a brief survey. International Journal of System Assurance Engineering and Management 4(2): 101–117.

Ram, M. and M. Manglik. 2016. Performance evaluation of a multi-state system covering imperfect fault coverage. Communications in Statistics-Simulation and Computation 45(9): 3259–3280.

Ram, M. and S.B. Singh. 2010. Analysis of a complex system with common cause failure and two types of repair facilities with different distributions in failure. International Journal of Reliability and Safety 4(4): 381–392.

Ram, M., S.B. Singh and V.V. Singh. 2013. Stochastic analysis of a standby system with waiting repair strategy. IEEE Transactions on Systems, Man, and Cybernetics: Systems 43(3): 698–707.

Salehi, E.T., M. Asadi and S. Eryılmaz. 2011. Reliability analysis of consecutive *k*-out-of-*n* systems with non-identical components lifetimes. Journal of Statistical Planning and Inference 141(8): 2920–2932.

Samrout, M., F. Yalaoui, E. Châtelet and N. Chebbo. 2005. New methods to minimize the preventive maintenance cost of series–parallel systems using ant colony optimization. Reliability Engineering and System Safety 89(3): 346–354.

Sharma, S., S.B. Pandey and S.B. Singh. 2009. Reliability and cost analysis of utility company website using middleware solution by mathematical modeling. Computer Modeling and New Technologies 13(2): 7–15.

Singh, V.V., M. Ram and D.K. Rawal. 2013. Cost analysis of an engineering system involving sub-systems in series configuration. IEEE Transactions on Automation Science and Engineering 10(4): 1124–1130.

Wang, K.-H., T.-C. Yen and Y.-C. Fang. 2012. Comparison of availability between two systems with warm standby units and different imperfect coverage. Quality Technology and Quantitative Management 9(3): 265–282.

Wang, W. and J. Loman. 2002. Reliability/availability of *k*-out-of-*n* system with *m* cold standby units. pp. 450–455. *In*: IEEE Reliability and Maintainability Symposium, 2002. Proceedings. Annual.

Wu, W., Y. Tang, M. Yu and Y. Jiang. 2014. Reliability analysis of a *k*-out-of-*n*:G repairable system with single vacation. Applied Mathematical Modelling 38(24): 6075–6097.

Yusuf, I. and N. Hussaini. 2012. Evaluation of reliability and availability characteristics of 2-out of-3 standby system under a perfect repair condition. American Journal of Mathematics and Statistics 2(5): 114–119.

Yusuf, I. and F.S. Koki. 2013. Evaluation of reliability and availability characteristics of a repairable system with active parallel units. Open Journal of Applied Sciences 3(05): 337.

Reliability Analysis of Three-stage Weighted 4-out-of-*n*:F System Subject to Possibility of Degradation after Repair with Inspection

Beena Nailwal[1*], Bhagawati Prasad Joshi[2] and S.B. Singh[1]

[1] G.B. Pant University of Agriculture and Technology, Pantnagar, India
[2] Seemant Institute of Technology, Pithoragarh, India

1. Introduction

Many researchers (including Dhillon 1992, Murari and Goyal 1984, Singh 1989, Srinivasan and Gopalan 1973) have studied the different types of systems under the assumption that the system becomes as good as new after repair. In real life situations, however, it is not always possible to do so, since the working capacity of a repaired unit depends, more or less, on the repair mechanism exercised as well as the age of the system. We can observe sometimes that if a failed unit is not being repaired by an expert server then the unit does not work as new and becomes degraded with increased chances of failure. Mokaddis et al. (1997) have analyzed a two-unit warm standby system subject to degradation. Malik et al. (2008) have discussed a system with inspection subject to degradation. Chander and Singh (2009), Kumar et al. (2010), and Promila et al. (2010) have studied the different systems subject to degradation after repair. In their study, they have assumed that the degraded unit after failure is not repaired completely. However, in practical life, there may be situations when the degraded failed unit does not work as good as new after repair or the repair of the degraded failed unit is neither possible nor economical due

*Corresponding author: bn4jan@gmail.com

to excessive use as well as high cost of maintenance. In such a situation, inspection plays an important role in deciding whether the unit that failed after degradation (degraded failed) can be operable as new after repair, or not. If the degraded failed unit does not work with full working capacity after repair, then it is advisable to replace it with new one in order to avoid unnecessary expenses on repair. Researchers (Nailwal and Singh 2011, 2012a, b, 2017, Ram, Singh and Singh 2013, Ram and Kumar 2014a, b, Manglik and Ram 2015, Kumar and Singh 2016, etc.) have discussed and studied different systems under different assumptions, and estimated their respective reliability parameters.

The different types of system configurations, such as *k*-out-of-*n* (Coit and Liu 2000), circular consecutive systems (Yam et al. 2003), etc., have been studied in the past. In most modern systems, the components have different contributions to the system which are defined as weights. Wu and Chen (1994) generalized the *k*-out-of-*n* system models into the weighted *k*-out-of-*n* models. A weighted *k*-out-of-*n*:G (F) system which has *n* components, each with their own positive integer weight ($w_i > 0$ for $i = 1, 2, ..., n$), such that the system is good (failed) if the total weight of good (failed) components is at least *k*, a prespecified value. Since *k* is a weight, it may be larger than *n* because they have different measuring units. The weight of the system is the sum of the weights of all working (failed) components. When a component has failed, its contribution to the system is zero. The *k*-out-of-*n*:G system is a special case of the weighted-*k*-out-of-*n*:G system in which the weight of each component is 1. The weight here means the contribution of the component in a system. In the weighted system, the component with higher contribution to the system has a higher weight and the component with lower contribution to the system has a lower weight. For example, a power generator with a higher output capacity has a larger weight. A lighting system having a fixed number of light bulbs with different performance characteristics such as different wattage and illumination power is a weighted system. When a certain level of illumination is deemed minimally acceptable, one can assume this value as the threshold value beyond which the system is defined as failed. Further, many practical components and systems have more than two different performance levels. For example, a power generator in a power station can work at full capacity, when there are no failures at all. Certain types of failures, such as catastrophic failure, can cause the generator to fail completely, while other failures, such as abnormal weather conditions or human error, will lead to the generator working at a reduced capacity. A component with multiple failure modes can also be considered to be a multistate component. Many systems, such as series-parallel and *k*-out-of-*n*, have been extended from two state cases to the multi state cases. Chen and Yang (2005) studied the reliability of two-stage weighted *k*-out-of-*n* systems with components in common.

Keeping the above facts in mind, this chapter has been designed to analyze the reliability measures and sensitivity analysis of three-stage weighted 4-out-of-n:F system, incorporating the concepts of degradation and inspection. The investigated system consists of two sub-systems, namely A and B, each having $n/2$ unit. The weight vector associated with sub-system A is w_A = (2, 2, ..., 2) and with the sub-system B is w_B = (3, 3, ..., 3). The system will fail completely if out of n units, the weight of failed units is 4. When the system has failed completely, it goes for repair where we have two possibilities, i.e. either the system becomes as good as new or degraded (not working with full efficiency) after repair. It is further assumed that, after degradation, the configuration of the system changes and becomes weighted 2-out-of-n:F. In this case, the system will be in failed state if out of n units, the weight of failed units is 2. There is a single repairman who plays the dual role of inspection and repair and is always available with the system. The repairman inspects the degraded system when it fails in order to assess the feasibility of repair. If there is no feasibility of repair, then the system is replaced by new one. We have used Gumbel-Hougaard family of copula (Nelson 2006, Sen 2003) to find joint distribution of repairs whenever both the sub-systems A and B are being repaired simultaneously with different repair rates. Failure rates are assumed to be constant in general, whereas the repairs and inspection follow general distribution in all the cases. The expressions for performance measures, such as transition state probabilities, asymptotic behavior, reliability, availability, MTTF, cost effectiveness and sensitivity analysis, are derived using supplementary variable technique, Laplace transformation and copula methodology. Graphs are plotted in order to depict the behavior of reliability measures for a particular case. Figures 1 and 2 represent the Block diagram of investigated system and transition state diagram respectively. Table 1 represents the state specification of the system.

2. Assumptions

1. Initially the system is in a perfectly operating state.
2. The considered system is 4-out-of-n:F having two sub-systems, namely A and B.
3. Each sub-system has $n/2$ units of weights 2 and 3, respectively.
4. The system will be in a failed state if out of n units, the weight of the failed units is 4.
5. There are two possibilities at the time of complete failure: (i) system works as good as new after repair and (ii) system does not work as good as new after repair and, therefore, becomes degraded.
6. When the system goes to degraded state, its configuration becomes weighted 2-out-of-n:F i.e. the system will fail if out of n units, the weight of failed unit(s) is 2.

7. There is single repairman who is always available for inspection and repair.
8. The repairman inspects the degraded system when it fails in order to assess the feasibility of repair. If the system cannot be repaired as good as new, it is replaced by new one.
9. The joint probability distribution of repairs when two failed subsystems A and B are repaired with different repair rates is obtained by Gumbel-Hougaard family of copula.

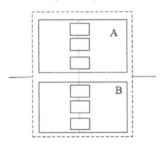

Fig. 1: Block diagram of investigated system

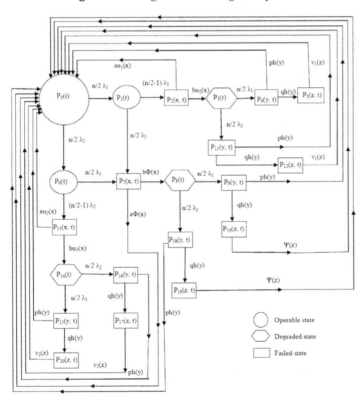

Fig. 2: Transition diagram

Table 1: State specification chart

States	System: Number of good units		Weight of failed units of system	System state
	Subsystem A	Subsystem B		
S_0	$n/2$	$n/2$	0	O
S_1	$(n/2)-1$	$n/2$	2	O
S_2	$(n/2)-2$	$n/2$	4	F_r
S_3	$n/2$	$n/2$	0	D
S_4	$(n/2)-1$	$n/2$	2	F_i
S_5	$(n/2)-1$	$n/2$	2	F_{dr}
S_6	$n/2$	$(n/2)-1$	3	O
S_7	$(n/2)-1$	$(n/2)-1$	5	F_r
S_8	$n/2$	$n/2$	0	D
S_9	$(n/2)-1$	$n/2$	2	F_i
S_{10}	$(n/2)-1$	$n/2$	2	F_{dr}
S_{11}	$n/2$	$(n/2)-1$	3	F_i
S_{12}	$n/2$	$(n/2)-1$	3	F_{dr}
S_{13}	$n/2$	$(n/2)-2$	6	F_r
S_{14}	$n/2$	$n/2$	0	D
S_{15}	$(n/2)-1$	$n/2$	2	F_i
S_{16}	$n/2$	$(n/2)-1$	3	F_i
S_{17}	$n/2$	$(n/2)-1$	3	F_{dr}
S_{18}	$n/2$	$(n/2)-1$	3	F_i
S_{19}	$n/2$	$(n/2)-1$	3	F_{dr}
S_{20}	$(n/2)-1$	$n/2$	2	F_{dr}

O: Operational state, F_r: Failed under repair, F_i: Failed under inspection,
D: Degraded after repair, F_{dr}: Failed under repair after degradation

3. Nomenclatures

λ_1	:	Failure rate of a sub-system A.
λ_2	:	Failure rate of a sub-system B.
$u_1(x)$	:	Repair rate of sub-system A.
$u_2(x)$	:	Repair rate of sub-system B.
$v_1(z)$	:	Repair rate of degraded sub-system A.
$v_2(z)$	:	Repair rate of degraded sub-system B.
$h(y)$	:	Inspection rate.
$\phi(x)$	:	Coupled repair rate when two sub-systems A and B are repaired.

$\psi(z)$	:	Coupled repair rate when two degraded sub-systems A and B are repaired.
x	:	Elapsed repair time.
y	:	Elapsed inspection time.
z	:	Elapsed repair time after degradation.
a/b	:	Probability that the system becomes as good as new/degraded after repair.
q/p	:	Probability that the degraded system works with full efficiency/ does not work with full capacity after repair.
$P_i(t)$	:	Probability that the system is in S_i state at instant t for $i = 1$ to 20.
$\overline{P}_i(s)$	:	Laplace transform of $P_i(t)$.
$P_i(x, t)$	:	Probability density function that at time t the system is in failed state S_i and the system is under repair, elapsed repair time is x.
$E_p(t)$	:	Expected profit during the interval $(0, t]$.
K_1, K_2	:	Revenue per unit time and service cost per unit time respectively.

$$S_\eta(x) \quad : \quad \eta(x)\exp\left(-\int_0^x \eta(x)dx\right)$$

$$\overline{S}_\eta(x) \quad : \quad \text{Laplace transform of } S_\eta(x) = \int_0^\infty \eta(x)\exp\left(-sx - \int_0^x \eta(x)dx\right)dx$$

If $a_1 = u_1(x)$, $a_2 = u_2(x)$, then the expression for the joint probability according with the help of Gumbel-Hougaard family of copula is given by

$$C_\theta(a_1 \cdot a_2) = \exp[-\{(-\log u_1(x))^\theta + (-\log u_2(x))^\theta\}^{1/\theta}]$$

4. Formulation of mathematical model

By elementary probability and continuity arguments, the set of integro-differential equations corresponding to the system is given by

$$\left[\frac{d}{dt} + \frac{n}{2}\lambda_1 + \frac{n}{2}\lambda_2\right]P_0(t) = \int_0^\infty au_1(x)P_2(x, t)dx + \int_0^\infty ph(y)P_4(y, t)dy$$

$$+ \int_0^\infty v_1(z)P_5(z, t)dz + \int_0^\infty ph(y)P_{11}(y, t)dy$$

$$+ \int_0^\infty v_1(z)P_{12}(z, t)dz + \int_0^\infty ph(y)P_9(y, t)dy + \int_0^\infty \psi(z)P_{10}(z, t)dz + \int_0^\infty \psi(z)P_{19}(z, t)dz$$

$$+ \int_0^\infty au_2(x)P_{13}(x, t)dx + \int_0^\infty ph(y)P_{15}(y, t)dy + \int_0^\infty v_2(z)P_{20}(z, t)dz + \int_0^\infty v_2(z)P_{17}(z, t)dz$$

$$+ \int_0^\infty ph(y)P_{16}(y, t)dy + \int_0^\infty a\phi(x)P_7(x, t)dx + \int_0^\infty ph(y)P_{18}(y, t)dy \tag{1}$$

$$\left[\frac{d}{dt} + \left(\frac{n}{2} - 1\right)\lambda_1 + \frac{n}{2}\lambda_2\right]P_1(t) = \frac{n}{2}\lambda_1 P_0(t) \tag{2}$$

$$\left[\frac{\partial}{\partial t} + \frac{\partial}{\partial x} + au_1(x) + bu_1(x)\right]P_2(x, t) = 0 \tag{3}$$

$$\left[\frac{d}{dt} + \frac{n}{2}\lambda_1 + \frac{n}{2}\lambda_2\right]P_3(t) = b\int_0^\infty u_1(x)P_2(x, t)dx \tag{4}$$

$$\left[\frac{\partial}{\partial t} + \frac{\partial}{\partial y} + ph(y) + qh(y)\right]P_4(y, t) = 0 \tag{5}$$

$$\left[\frac{\partial}{\partial t} + \frac{\partial}{\partial z} + v_1(z)\right]P_5(z, t) = 0 \tag{6}$$

$$\left[\frac{d}{dt} + \left(\frac{n}{2} - 1\right)\lambda_2 + \frac{n}{2}\lambda_1\right]P_6(t) = \frac{n}{2}\lambda_2 P_0(t) \tag{7}$$

$$\left[\frac{\partial}{\partial t} + \frac{\partial}{\partial x} + a\phi(x) + b\phi(x)\right]P_7(x, t) = 0 \tag{8}$$

$$\left[\frac{d}{dt} + \frac{n}{2}\lambda_1 + \frac{n}{2}\lambda_2\right]P_8(t) = b\int_0^\infty \phi(x)P_7(x, t)dx \tag{9}$$

$$\left[\frac{\partial}{\partial t} + \frac{\partial}{\partial y} + ph(y) + qh(y)\right]P_9(y, t) = 0 \tag{10}$$

$$\left[\frac{\partial}{\partial t} + \frac{\partial}{\partial z} + \psi(z)\right]P_{10}(z, t) = 0 \tag{11}$$

$$\left[\frac{\partial}{\partial t} + \frac{\partial}{\partial y} + ph(y) + qh(y)\right]P_{11}(y, t) = 0 \tag{12}$$

$$\left[\frac{\partial}{\partial t} + \frac{\partial}{\partial z} + v_1(z)\right]P_{12}(z, t) = 0 \tag{13}$$

$$\left[\frac{\partial}{\partial t} + \frac{\partial}{\partial x} + av_2(x) + bv_2(x)\right]P_{13}(x, t) = 0 \tag{14}$$

$$\left[\frac{d}{dt} + \frac{n}{2}\lambda_1 + \frac{n}{2}\lambda_2\right]P_{14}(t) = b\int_0^\infty u_2(x)P_{13}(x, t)dx \tag{15}$$

$$\left[\frac{\partial}{\partial t} + \frac{\partial}{\partial y} + ph(y) + qh(y)\right]P_{15}(y, t) = 0 \tag{16}$$

$$\left[\frac{\partial}{\partial t} + \frac{\partial}{\partial y} + ph(y) + qh(y)\right]P_{16}(y, t) = 0 \tag{17}$$

$$\left[\frac{\partial}{\partial t} + \frac{\partial}{\partial z} + v_2(z)\right]P_{17}(z, t) = 0 \tag{18}$$

$$\left[\frac{\partial}{\partial t} + \frac{\partial}{\partial y} + ph(y) + qh(y)\right]P_{18}(y, t) = 0 \tag{19}$$

$$\left[\frac{\partial}{\partial t} + \frac{\partial}{\partial z} + \psi(z)\right]P_{19}(z, t) = 0 \tag{20}$$

$$\left[\frac{\partial}{\partial t} + \frac{\partial}{\partial z} + v_2(z)\right]P_{20}(z, t) = 0 \tag{21}$$

Boundary conditions:

$$P_2(0, t) = \left(\frac{n}{2} - 1\right)\lambda_1 P_1(t) \tag{22}$$

$$P_4(0, t) = \frac{n}{2}\lambda_1 P_3(t) \tag{23}$$

$$P_5(0, t) = \int_0^\infty qh(y)P_4(y, t)dy \tag{24}$$

$$P_7(0, t) = \frac{n}{2}\lambda_1 P_6(t) + \frac{n}{2}\lambda_2 P_1(t) \tag{25}$$

$$P_9(0, t) = \frac{n}{2}\lambda_1 P_8(t) \tag{26}$$

$$P_{10}(0, t) = \int_0^\infty qh(y)P_9(y, t)dy \tag{27}$$

$$P_{11}(0, t) = \frac{n}{2}\lambda_2 P_3(t) \tag{28}$$

$$P_{12}(0, t) = \int_0^\infty qh(y)P_{11}(y, t)dy \tag{29}$$

$$P_{13}(0, t) = \left(\frac{n}{2} - 1\right)\lambda_2 P_6(t) \tag{30}$$

$$P_{15}(0, t) = \frac{n}{2}\lambda_1 P_{14}(t) \tag{31}$$

$$P_{16}(0, t) = \frac{n}{2}\lambda_2 P_{14}(t) \tag{32}$$

$$P_{17}(0, t) = \int_0^\infty qh(y)P_{16}(y, t)dy \tag{33}$$

$$P_{18}(0, t) = \frac{n}{2}\lambda_2 P_8(t) \tag{34}$$

$$P_{19}(0, t) = \int_0^\infty qh(y)P_{18}(y, t)dy \tag{35}$$

$$P_{20}(0, t) = \int_0^\infty qh(y)P_{15}(y, t)dy \tag{36}$$

Initial condition:
$P_0(t) = 1$ at $t = 0$ and all other probabilities are zero initially.

5. Solution of the model

Taking Laplace transformation of (1) to (36), and on further simplification, one can obtain transition state probabilities of the system as:

$$\overline{P}_0(s) = \frac{1}{D(s)} \tag{37}$$

$$\overline{P}_1(s) = \frac{n}{2} \times \frac{\lambda_1 \overline{P}_0(s)}{\left[s + \left(\frac{n}{2} - 1\right)\lambda_1 + \frac{n}{2}\lambda_2\right]} \tag{38}$$

$$\overline{P}_2(s) = \left(\frac{n}{2} - 1\right) \times \frac{n}{2} \times \frac{\lambda_1^2 \overline{P}_0(s)}{\left[s + \left(\frac{n}{2} - 1\right)\lambda_1 + \frac{n}{2}\lambda_2\right]} \times \left[\frac{1 - \overline{S}_{u_1}(s)}{s}\right] \tag{39}$$

$$\overline{P}_3(s) = \left(\frac{n}{2} - 1\right) \times \frac{n}{2} \times \frac{b\lambda_1^2 \overline{P}_0(s)}{\left[s + \frac{n}{2}\lambda_1 + \frac{n}{2}\lambda_2\right]\left[s + \left(\frac{n}{2} - 1\right)\lambda_1 + \frac{n}{2}\lambda_2\right]} \times \overline{S}_{u_1}(s) \tag{40}$$

$$\overline{P}_4(s) = \frac{n^2}{4} \times \left(\frac{n}{2}-1\right) \times \frac{b\lambda_1{}^3 \overline{P}_0(s)\overline{S}_{u_1}(s)}{\left[s+\frac{n}{2}\lambda_1+\frac{n}{2}\lambda_2\right]\left[s+\left(\frac{n}{2}-1\right)\lambda_1+\frac{n}{2}\lambda_2\right]} \times \left[\frac{1-\overline{S}_h(s)}{s}\right]$$

(41)

$$\overline{P}_5(s) = \frac{n^2}{4} \times \left(\frac{n}{2}-1\right) \times \frac{qb\lambda_1{}^3 \overline{P}_0(s)\overline{S}_{u_1}(s)\overline{S}_h(s)}{\left[s+\frac{n}{2}\lambda_1+\frac{n}{2}\lambda_2\right]\left[s+\left(\frac{n}{2}-1\right)\lambda_1+\frac{n}{2}\lambda_2\right]} \times \left[\frac{1-\overline{S}_{v_1}(s)}{s}\right]$$

(42)

$$\overline{P}_6(s) = \frac{n}{2} \times \frac{\lambda_2 \overline{P}_0(s)}{\left[s+\left(\frac{n}{2}-1\right)\lambda_2+\frac{n}{2}\lambda_1\right]}$$

(43)

$$\overline{P}_7(s) = \frac{n^2}{4} \times \frac{\lambda_1\lambda_2 \overline{P}_0(s)}{\left[s+\left(\frac{n}{2}-1\right)\lambda_2+\frac{n}{2}\lambda_1\right]} \times \frac{\left[2s+(n-1)\lambda_1+(n-1)\lambda_2\right]}{\left[s+\left(\frac{n}{2}-1\right)\lambda_1+\frac{n}{2}\lambda_2\right]}$$
$$\times \left[\frac{1-\overline{S}_\phi(s)}{s}\right]$$

(44)

$$\overline{P}_8(s) = \frac{b\overline{S}_\phi(s)}{\left[s+\frac{n}{2}\lambda_1+\frac{n}{2}\lambda_2\right]} \times \frac{n^2}{4} \times \frac{\lambda_1\lambda_2 \overline{P}_0(s)}{\left[s+\left(\frac{n}{2}-1\right)\lambda_2+\frac{n}{2}\lambda_1\right]}$$
$$\times \frac{\left[2s+(n-1)\lambda_1+(n-1)\lambda_2\right]}{\left[s+\left(\frac{n}{2}-1\right)\lambda_1+\frac{n}{2}\lambda_2\right]}$$

(45)

$$\overline{P}_9(s) = \frac{b\overline{S}_\phi(s)}{\left[s+\frac{n}{2}\lambda_1+\frac{n}{2}\lambda_2\right]} \times \frac{n^3}{8} \times \frac{\lambda_1{}^2\lambda_2 \overline{P}_0(s)}{\left[s+\left(\frac{n}{2}-1\right)\lambda_2+\frac{n}{2}\lambda_1\right]}$$
$$\times \frac{\left[2s+(n-1)\lambda_1+(n-1)\lambda_2\right]}{\left[s+\left(\frac{n}{2}-1\right)\lambda_1+\frac{n}{2}\lambda_2\right]}\left[\frac{1-\overline{S}_h(s)}{s}\right]$$

(46)

$$\overline{P}_{10}(s) = \frac{b\overline{S}_\phi(s)}{\left[s+\dfrac{n}{2}\lambda_1+\dfrac{n}{2}\lambda_2\right]} \times \frac{n^3}{8} \times \frac{q\lambda_1{}^2\lambda_2\overline{P}_0(s)\overline{S}_h(s)}{\left[s+\left(\dfrac{n}{2}-1\right)\lambda_2+\dfrac{n}{2}\lambda_1\right]}$$

$$\times \frac{\left[2s+(n-1)\lambda_1+(n-1)\lambda_2\right]}{\left[s+\left(\dfrac{n}{2}-1\right)\lambda_1+\dfrac{n}{2}\lambda_2\right]}\left[\frac{1-\overline{S}_\psi(s)}{s}\right] \quad (47)$$

$$\overline{P}_{11}(s) = \left(\frac{n}{2}-1\right) \times \frac{n^2}{4} \times \frac{\lambda_2 b\lambda_1{}^2\overline{P}_0(s)}{\left[s+\dfrac{n}{2}\lambda_1+\dfrac{n}{2}\lambda_2\right]\left[s+\left(\dfrac{n}{2}-1\right)\lambda_1+\dfrac{n}{2}\lambda_2\right]}$$

$$\times \overline{S}_{u_1}(s)\left[\frac{1-\overline{S}_h(s)}{s}\right] \quad (48)$$

$$\overline{P}_{12}(s) = \left(\frac{n}{2}-1\right) \times \frac{n^2}{4} \times \frac{q\lambda_2 b\lambda_1{}^2\overline{P}_0(s)\overline{S}_{u_1}(s)\overline{S}_h(s)}{\left[s+\dfrac{n}{2}\lambda_1+\dfrac{n}{2}\lambda_2\right]\left[s+\left(\dfrac{n}{2}-1\right)\lambda_1+\dfrac{n}{2}\lambda_2\right]}$$

$$\times \left[\frac{1-\overline{S}_{v_1}(s)}{s}\right] \quad (49)$$

$$\overline{P}_{13}(s) = \left(\frac{n}{2}-1\right) \times \frac{n}{2} \times \frac{\lambda_2{}^2\overline{P}_0(s)}{\left[s+\left(\dfrac{n}{2}-1\right)\lambda_2+\dfrac{n}{2}\lambda_1\right]}\left[\frac{1-\overline{S}_{u_2}(s)}{s}\right] \quad (50)$$

$$\overline{P}_{14}(s) = \left(\frac{n}{2}-1\right) \times \frac{n}{2} \times \frac{b\lambda_2{}^2\overline{P}_0(s)\overline{S}_{u_2}(s)}{\left[s+\dfrac{n}{2}\lambda_1+\dfrac{n}{2}\lambda_2\right]\left[s+\left(\dfrac{n}{2}-1\right)\lambda_2+\dfrac{n}{2}\lambda_1\right]} \quad (51)$$

$$\overline{P}_{15}(s) = \left(\frac{n}{2}-1\right) \times \frac{n^2}{4} \times \frac{\lambda_1 b\lambda_2{}^2\overline{P}_0(s)\overline{S}_{u_2}(s)}{\left[s+\dfrac{n}{2}\lambda_1+\dfrac{n}{2}\lambda_2\right]\left[s+\left(\dfrac{n}{2}-1\right)\lambda_2+\dfrac{n}{2}\lambda_1\right]}\left[\frac{1-\overline{S}_h(s)}{s}\right]$$

$$(52)$$

$$\overline{P}_{16}(s) = \left(\frac{n}{2}-1\right) \times \frac{n^2}{4} \times \frac{b\lambda_2{}^3\overline{P}_0(s)\overline{S}_{u_2}(s)}{\left[s+\dfrac{n}{2}\lambda_1+\dfrac{n}{2}\lambda_2\right]\left[s+\left(\dfrac{n}{2}-1\right)\lambda_2+\dfrac{n}{2}\lambda_1\right]}\left[\frac{1-\overline{S}_h(s)}{s}\right]$$

$$(53)$$

$$\overline{P}_{17}(s) = \left(\frac{n}{2} - 1\right) \times \frac{n^2}{4} \times \frac{qb\lambda_2{}^3 \overline{P}_0(s)\overline{S}_{u_2}(s)\overline{S}_h(s)}{\left[s + \frac{n}{2}\lambda_1 + \frac{n}{2}\lambda_2\right]\left[s + \left(\frac{n}{2} - 1\right)\lambda_2 + \frac{n}{2}\lambda_1\right]} \left[\frac{1 - \overline{S}_{v_2}(s)}{s}\right]$$

$$(54)$$

$$\overline{P}_{18}(s) = \frac{b\overline{S}_\phi(s)}{\left[s + \frac{n}{2}\lambda_1 + \frac{n}{2}\lambda_2\right]} \times \frac{n^3}{8} \times \frac{\lambda_1\lambda_2{}^2 \overline{P}_0(s)}{\left[s + \left(\frac{n}{2} - 1\right)\lambda_2 + \frac{n}{2}\lambda_1\right]}$$

$$\times \frac{\left[2s + (n-1)\lambda_1 + (n-1)\lambda_2\right]}{\left[s + \left(\frac{n}{2} - 1\right)\lambda_1 + \frac{n}{2}\lambda_2\right]} \left[\frac{1 - \overline{S}_h(s)}{s}\right] \quad (55)$$

$$\overline{P}_{19}(s) = \frac{b\overline{S}_\phi(s)}{\left[s + \frac{n}{2}\lambda_1 + \frac{n}{2}\lambda_2\right]} \times \frac{n^3}{8} \times \frac{q\lambda_1\lambda_2{}^2 \overline{P}_0(s)\overline{S}_h(s)}{\left[s + \left(\frac{n}{2} - 1\right)\lambda_2 + \frac{n}{2}\lambda_1\right]}$$

$$\times \frac{\left[2s + (n-1)\lambda_1 + (n-1)\lambda_2\right]}{\left[s + \left(\frac{n}{2} - 1\right)\lambda_1 + \frac{n}{2}\lambda_2\right]} \left[\frac{1 - \overline{S}_\psi(s)}{s}\right] \quad (56)$$

$$\overline{P}_{20}(s) = \left(\frac{n}{2} - 1\right) \times \frac{n^2}{4} \times \frac{q\lambda_1 b\lambda_2{}^2 \overline{P}_0(s)\overline{S}_{u_2}(s)\overline{S}_h(s)}{\left[s + \frac{n}{2}\lambda_1 + \frac{n}{2}\lambda_2\right]\left[s + \left(\frac{n}{2} - 1\right)\lambda_2 + \frac{n}{2}\lambda_1\right]}$$

$$\times \left[\frac{1 - \overline{S}_{v_2}(s)}{s}\right] \quad (57)$$

where

$$D(s) = \left(s + \frac{n}{2}\lambda_1 + \frac{n}{2}\lambda_2\right) - \left(\frac{n}{2} - 1\right) \times \frac{a\frac{n}{2}\lambda_1{}^2 \overline{S}_{u_1}(s)}{\left[s + \left(\frac{n}{2} - 1\right)\lambda_1 + \frac{n}{2}\lambda_2\right]}$$

$$- \frac{b\lambda_1{}^3 p\overline{S}_{u_1}(s)\overline{S}_h(s)}{\left[s + \frac{n}{2}\lambda_1 + \frac{n}{2}\lambda_2\right]\left[s + \left(\frac{n}{2} - 1\right)\lambda_1 + \frac{n}{2}\lambda_2\right]} \times \frac{n^2}{4}$$

$$\left(\frac{n}{2}-1\right)-\left(\frac{n}{2}-1\right)\times\frac{n^2}{4}\times\frac{qb\lambda_1{}^3\,\bar{S}_{u_1}(s)\bar{S}_{v_1}(s)\bar{S}_h(s)}{\left[s+\frac{n}{2}\lambda_1+\frac{n}{2}\lambda_2\right]\left[s+\left(\frac{n}{2}-1\right)\lambda_1+\frac{n}{2}\lambda_2\right]}$$

$$-\frac{p\lambda_2 b\lambda_1{}^2\,\bar{S}_{u_1}(s)\bar{S}_h(s)}{\left[s+\frac{n}{2}\lambda_1+\frac{n}{2}\lambda_2\right]\left[s+\left(\frac{n}{2}-1\right)\lambda_1+\frac{n}{2}\lambda_2\right]}$$

$$\left(\frac{n}{2}-1\right)\times\frac{n^2}{4}-\left(\frac{n}{2}-1\right)\times\frac{n^2}{4}\times\frac{q\lambda_2 b\lambda_1{}^2\,\bar{S}_{u_1}(s)\bar{S}_h(s)\bar{S}_{v_1}(s)}{\left[s+\frac{n}{2}\lambda_1+\frac{n}{2}\lambda_2\right]\left[s+\left(\frac{n}{2}-1\right)\lambda_1+\frac{n}{2}\lambda_2\right]}$$

$$-\frac{p\left[2s+(n-1)\lambda_1+(n-1)\lambda_2\right]}{\left[s+\left(\frac{n}{2}-1\right)\lambda_1+\frac{n}{2}\lambda_2\right]}\times\frac{n^3}{8}$$

$$\frac{b\bar{S}_\phi(s)}{\left[s+\frac{n}{2}\lambda_1+\frac{n}{2}\lambda_2\right]}\times\frac{\lambda_1{}^2\lambda_2\,\bar{S}_h(s)}{\left[s+\left(\frac{n}{2}-1\right)\lambda_2+\frac{n}{2}\lambda_1\right]}-\frac{q\left[2s+(n-1)\lambda_1+(n-1)\lambda_2\right]}{\left[s+\left(\frac{n}{2}-1\right)\lambda_1+\frac{n}{2}\lambda_2\right]}$$

$$\times\frac{n^3}{8}\frac{b\bar{S}_\phi(s)}{\left[s+\frac{n}{2}\lambda_1+\frac{n}{2}\lambda_2\right]}$$

$$\times\frac{\lambda_1{}^2\lambda_2\,\bar{S}_h(s)\bar{S}_\psi(s)}{\left[s+\left(\frac{n}{2}-1\right)\lambda_2+\frac{n}{2}\lambda_1\right]}-\frac{q\left[2s+(n-1)\lambda_1+(n-1)\lambda_2\right]}{\left[s+\left(\frac{n}{2}-1\right)\lambda_1+\frac{n}{2}\lambda_2\right]}\times\frac{n^3}{8}$$

$$\times\frac{b\bar{S}_\phi(s)}{\left[s+\frac{n}{2}\lambda_1+\frac{n}{2}\lambda_2\right]}\times\frac{\lambda_1\lambda_2{}^2\,\bar{S}_h(s)\bar{S}_\psi(s)}{\left[s+\left(\frac{n}{2}-1\right)\lambda_2+\frac{n}{2}\lambda_1\right]}$$

$$-\left(\frac{n}{2}-1\right)\times\frac{n}{2}\times\frac{a\lambda_2{}^2\,\bar{S}_{u_2}(s)}{\left[s+\left(\frac{n}{2}-1\right)\lambda_2+\frac{n}{2}\lambda_1\right]}-\left(\frac{n}{2}-1\right)\times\frac{n^2}{4}$$

$$\times\frac{pb\lambda_1\lambda_2{}^2\,\bar{P}_0(s)\bar{S}_{u_2}(s)\bar{S}_h(s)}{\left[s+\frac{n}{2}\lambda_1+\frac{n}{2}\lambda_2\right]\left[s+\left(\frac{n}{2}-1\right)\lambda_2+\frac{n}{2}\lambda_1\right]}-\left(\frac{n}{2}-1\right)$$

$$\times \frac{n^2}{4} \times \frac{qb\lambda_1\lambda_2{}^2 \overline{P}_0(s)\overline{S}_{u_2}(s)\overline{S}_h(s)\overline{S}_{v_2}(s)}{\left[s+\dfrac{n}{2}\lambda_1+\dfrac{n}{2}\lambda_2\right]\left[s+\left(\dfrac{n}{2}-1\right)\lambda_2+\dfrac{n}{2}\lambda_1\right]} - \left(\dfrac{n}{2}-1\right)\times\frac{n^2}{4}$$

$$\times \frac{qb\lambda_2{}^3 \overline{P}_0(s)\overline{S}_{u_2}(s)\overline{S}_h(s)\overline{S}_{v_2}(s)}{\left[s+\dfrac{n}{2}\lambda_1+\dfrac{n}{2}\lambda_2\right]\left[s+\left(\dfrac{n}{2}-1\right)\lambda_2+\dfrac{n}{2}\lambda_1\right]} \times\left(\dfrac{n}{2}-1\right)\times\frac{n^2}{4}$$

$$\times \frac{pb\lambda_2{}^3 \overline{P}_0(s)\overline{S}_{u_2}(s)\overline{S}_h(s)}{\left[s+\dfrac{n}{2}\lambda_1+\dfrac{n}{2}\lambda_2\right]\left[s+\left(\dfrac{n}{2}-1\right)\lambda_2+\dfrac{n}{2}\lambda_1\right]} - \frac{n^2}{4}\times\frac{a\lambda_1\lambda_2\overline{S}_\phi(s)}{\left[s+\left(\dfrac{n}{2}-1\right)\lambda_2+\dfrac{n}{2}\lambda_1\right]}$$

$$\times \frac{\left[2s+(n-1)\lambda_1+(n-1)\lambda_2\right]}{\left[s+\left(\dfrac{n}{2}-1\right)\lambda_1+\dfrac{n}{2}\lambda_2\right]} - \frac{pb\overline{S}_\phi(s)}{\left[s+\dfrac{n}{2}\lambda_1+\dfrac{n}{2}\lambda_2\right]}\times\frac{n^3}{8}$$

$$\times \frac{\lambda_1\lambda_2{}^2\overline{S}_h(s)}{\left[s+\left(\dfrac{n}{2}-1\right)\lambda_2+\dfrac{n}{2}\lambda_1\right]} \times \frac{\left[2s+(n-1)\lambda_1+(n-1)\lambda_2\right]}{\left[s+\left(\dfrac{n}{2}-1\right)\lambda_1+\dfrac{n}{2}\lambda_2\right]}$$

$$\phi(x) = \exp[-\{(-\log u_1(x))^\theta + (-\log u_2(x))^\theta\}^{1/\theta}]$$

$$\psi(x) = \exp[-\{(-\log v_1(x))^\theta + (-\log v_2(x))^\theta\}^{1/\theta}]$$

The up and down state probabilities of the system are given by

$$\overline{P}_{up}(s) = \overline{P}_0(s) + \overline{P}_1(s) + \overline{P}_3(s) + \overline{P}_6(s) + \overline{P}_8(s) + \overline{P}_{14}(s) \qquad (58)$$

$$\overline{P}_{down}(s) = \overline{P}_2(s) + \overline{P}_4(s) + \overline{P}_5(s) + \overline{P}_7(s) + \overline{P}_9(s) + \overline{P}_{10}(s) + \overline{P}_{11}(s) + \overline{P}_{12}(s)$$
$$+ \overline{P}_{13}(s) + \overline{P}_{15}(s) + \overline{P}_{16}(s) + \overline{P}_{17}(s) + \overline{P}_{18}(s) + \overline{P}_{19}(s) + \overline{P}_{20}(s) \qquad (59)$$

From (61) and (62), we have

$$\overline{P}_{up}(s) + \overline{P}_{down}(s) = 1/s$$

6. Asymptotic behavior

The Able's lemma is expressed as

$$\lim_{s\to 0}\{s\overline{F}(s)\} = \lim_{t\to\infty} F(t)$$

Using this in equations (58) and (59), the expressions for the up and down steady state probabilities of the system take the form as given in the equations below:

$$P_{up} = \frac{1}{D(0)}\left[1 + \frac{n}{2} \times \frac{\lambda_1}{\left[\left(\frac{n}{2}-1\right)\lambda_1 + \frac{n}{2}\lambda_2\right]} + \left(\frac{n}{2}-1\right) \times \frac{n}{2} \right.$$

$$\times \frac{b\lambda_1^2}{\left[\frac{n}{2}\lambda_1 + \frac{n}{2}\lambda_2\right]\left[\left(\frac{n}{2}-1\right)\lambda_1 + \frac{n}{2}\lambda_2\right]} \times \overline{N}_{u_1} + \frac{n}{2} \times \frac{\lambda_2}{\left[\left(\frac{n}{2}-1\right)\lambda_2 + \frac{n}{2}\lambda_1\right]}$$

$$+ \frac{b\overline{N}_\phi}{\left[\frac{n}{2}\lambda_1 + \frac{n}{2}\lambda_2\right]} \times \frac{n^2}{4} \times \frac{\lambda_1\lambda_2}{\left[\left(\frac{n}{2}-1\right)\lambda_2 + \frac{n}{2}\lambda_1\right]} \times \frac{\left[(n-1)\lambda_1 + (n-1)\lambda_2\right]}{\left[\left(\frac{n}{2}-1\right)\lambda_1 + \frac{n}{2}\lambda_2\right]}$$

$$\left. + \left(\frac{n}{2}-1\right) \times \frac{n}{2} \times \frac{b\lambda_2^2\overline{N}_{u_2}}{\left[\frac{n}{2}\lambda_1 + \frac{n}{2}\lambda_2\right]\left[\left(\frac{n}{2}-1\right)\lambda_2 + \frac{n}{2}\lambda_1\right]} \right]$$

$$P_{down} = \frac{1}{D(0)}\left[\left(\frac{n}{2}-1\right) \times \frac{n}{2} \times \frac{\lambda_1^2}{\left[\left(\frac{n}{2}-1\right)\lambda_1 + \frac{n}{2}\lambda_2\right]} \times \overline{M}_{u_1} + \frac{n^2}{4} \times \left(\frac{n}{2}-1\right) \right.$$

$$\times \frac{b\lambda_1^3\overline{N}_{u_1}}{\left[\frac{n}{2}\lambda_1 + \frac{n}{2}\lambda_2\right]\left[\left(\frac{n}{2}-1\right)\lambda_1 + \frac{n}{2}\lambda_2\right]} \times \overline{M}_h + \frac{n^2}{4} \times \left(\frac{n}{2}-1\right)$$

$$\times \frac{qb\lambda_1^3\overline{N}_{u_1}\overline{N}_h}{\left[\frac{n}{2}\lambda_1 + \frac{n}{2}\lambda_2\right]\left[\left(\frac{n}{2}-1\right)\lambda_1 + \frac{n}{2}\lambda_2\right]} \times \overline{M}_{v_1} + \frac{n^2}{4}$$

$$\left. \times \frac{\lambda_1\lambda_2}{\left[\left(\frac{n}{2}-1\right)\lambda_2 + \frac{n}{2}\lambda_1\right]} \times \frac{\left[(n-1)\lambda_1 + (n-1)\lambda_2\right]}{\left[\left(\frac{n}{2}-1\right)\lambda_1 + \frac{n}{2}\lambda_2\right]} \times \overline{M}_\phi \right.$$

$$+\frac{b\overline{N}_\phi}{\left[\frac{n}{2}\lambda_1+\frac{n}{2}\lambda_2\right]}\times\frac{n^3}{8}\times\frac{\lambda_1^2\lambda_2}{\left[\left(\frac{n}{2}-1\right)\lambda_2+\frac{n}{2}\lambda_1\right]}$$

$$\times\frac{\left[(n-1)\lambda_1+(n-1)\lambda_2\right]}{\left[\left(\frac{n}{2}-1\right)\lambda_1+\frac{n}{2}\lambda_2\right]}\overline{M}_h+\frac{b\overline{N}_\phi}{\left[\frac{n}{2}\lambda_1+\frac{n}{2}\lambda_2\right]}\times\frac{n^3}{8}$$

$$\times\frac{q\lambda_1^2\lambda_2\overline{N}_h}{\left[\left(\frac{n}{2}-1\right)\lambda_2+\frac{n}{2}\lambda_1\right]}\times\frac{\left[(n-1)\lambda_1+(n-1)\lambda_2\right]}{\left[\left(\frac{n}{2}-1\right)\lambda_1+\frac{n}{2}\lambda_2\right]}$$

$$\times\overline{M}_\psi+\left(\frac{n}{2}-1\right)\times\frac{n^2}{4}\times\frac{\lambda_2 b\lambda_1^2}{\left[\frac{n}{2}\lambda_1+\frac{n}{2}\lambda_2\right]\left[\left(\frac{n}{2}-1\right)\lambda_1+\frac{n}{2}\lambda_2\right]}$$

$$\times\overline{N}_{u_1}\times\overline{M}_h+\left(\frac{n}{2}-1\right)\times\frac{q\lambda_2 b\lambda_1^2\overline{N}_{u_1}\overline{N}_h}{\left[\frac{n}{2}\lambda_1+\frac{n}{2}\lambda_2\right]\left[\left(\frac{n}{2}-1\right)\lambda_1+\frac{n}{2}\lambda_2\right]}$$

$$\times\overline{M}_{v_1}\frac{n^2}{4}+\left(\frac{n}{2}-1\right)\times\frac{n}{2}\times\frac{\lambda_2^2}{\left[\left(\frac{n}{2}-1\right)\lambda_2+\frac{n}{2}\lambda_1\right]}\times\overline{M}_{u_2}+\left(\frac{n}{2}-1\right)$$

$$\times\frac{\lambda_1 b\lambda_2^2\overline{N}_{u_2}}{\left[\frac{n}{2}\lambda_1+\frac{n}{2}\lambda_2\right]\left[\left(\frac{n}{2}-1\right)\lambda_2+\frac{n}{2}\lambda_1\right]}\times\overline{M}_h\times\frac{n^2}{4}+\left(\frac{n}{2}-1\right)$$

$$\times\frac{n^2}{4}\times\frac{b\lambda_2^3\overline{N}_{u_2}}{\left[\frac{n}{2}\lambda_1+\frac{n}{2}\lambda_2\right]\left[\left(\frac{n}{2}-1\right)\lambda_2+\frac{n}{2}\lambda_1\right]}\times\overline{M}_h$$

$$+\left(\frac{n}{2}-1\right)\times\frac{n^2}{4}\times\frac{qb\lambda_2^3\overline{N}_{u_2}\overline{N}_h}{\left[\frac{n}{2}\lambda_1+\frac{n}{2}\lambda_2\right]\left[\left(\frac{n}{2}-1\right)\lambda_2+\frac{n}{2}\lambda_1\right]}\overline{M}_{v_2}$$

$$+\frac{b\overline{N}_\phi}{\left[\frac{n}{2}\lambda_1+\frac{n}{2}\lambda_2\right]}\times\frac{n^3}{8}\times\frac{\lambda_1\lambda_2^2}{\left[\left(\frac{n}{2}-1\right)\lambda_2+\frac{n}{2}\lambda_1\right]}$$

$$\times \frac{\left[(n-1)\lambda_1 + (n-1)\lambda_2\right]}{\left[\left(\dfrac{n}{2}-1\right)\lambda_1 + \dfrac{n}{2}\lambda_2\right]} \times \overline{M}_h + \frac{b\overline{N}_\phi}{\left[\dfrac{n}{2}\lambda_1 + \dfrac{n}{2}\lambda_2\right]} \times \frac{n^3}{8}$$

$$\times \frac{q\lambda_1\lambda_2{}^2\overline{N}_h}{\left[\left(\dfrac{n}{2}-1\right)\lambda_2 + \dfrac{n}{2}\lambda_1\right]} \times \frac{\left[(n-1)\lambda_1 + (n-1)\lambda_2\right]}{\left[\left(\dfrac{n}{2}-1\right)\lambda_1 + \dfrac{n}{2}\lambda_2\right]}$$

$$\times \overline{M}_\psi + \left(\dfrac{n}{2}-1\right) \times \dfrac{n^2}{4} \times \frac{q\lambda_1 b\lambda_2^2 \overline{N}_{u_2}\overline{N}_h}{\left[\dfrac{n}{2}\lambda_1 + \dfrac{n}{2}\lambda_2\right]\left[\left(\dfrac{n}{2}-1\right)\lambda_2 + \dfrac{n}{2}\lambda_1\right]} \times \overline{M}_{v_2} \right]$$

where $D(0) = \lim\limits_{s\to 0} D(s)$, $\overline{M}_\phi = \lim\limits_{s\to 0} \dfrac{1-\overline{S}_\phi(s)}{s}$, $\overline{N}_\phi = \lim\limits_{s\to 0}\overline{S}_\phi(s)$.

7. Particular case

7.1 When repair follows exponential distribution

In this case the results can be derived by putting

$$\overline{S}_{u_1}(s) = \frac{u_1(x)}{s + u_1(x)}, \overline{S}_{u_2}(s) = \frac{u_2(x)}{s + u_2(x)}, \overline{S}_{v_1}(s) = \frac{v_1(x)}{s + v_1(x)}$$

$$\overline{S}_{v_2}(s) = \frac{v_2(x)}{s + v_2(x)}, \overline{S}_\psi(s) = \frac{\psi(x)}{s + \psi(x)}, \overline{S}_\phi(s) = \frac{\phi(x)}{s + \phi(x)} \tag{60}$$

in equations (58) and (59).

8. Numerical computation

To numerically examine the model, failure rates and some other parameters are kept fixed as

$$\lambda_1 = 0.01,\ \lambda_2 = 0.01,\ n = 6,\ \theta = 1,\ x = y = z = 1 \tag{61}$$

in order to obtain the reliability, availability and cost analysis of the system. Also let the repairs follow exponential distribution, i.e. equation (60) holds.

8.1 Reliability analysis

In order to obtain the reliability of the system, we take $u_1 = u_2 = v_1 = v_2 = \Phi = \psi = h = 0$ and the values mentioned in equation (61). Now, by setting $t =$

$0, 1, 2, 3, 4, 5, 6, 7, 8, 9, 10$, one can obtain Table 2 and, correspondingly, Fig. 3 which represents how reliability changes as time increases.

Table 2: Time vs. reliability

Time	Reliability
0	1
1	0.998553879
2	0.994422325
3	0.987896802
4	0.979245213
5	0.968713595
6	0.956527694
7	0.942894439
8	0.928003317
9	0.912027648
10	0.895125778

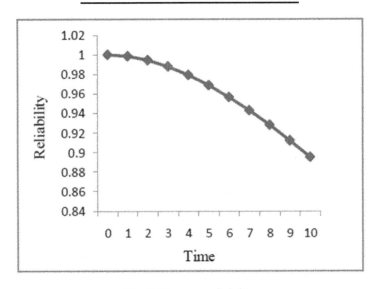

Fig. 3: Time vs. reliability

8.2 Availability analysis

To evaluate availability, we take all values given in equation (61) and set $u_1 = u_2 = v_1 = v_2 = \Phi = \psi = h = 1$, $p = 0.2$, $q = 0.8$, $a = 0.2$ and $b = 0.8$ in equation (58) and taking inverse Laplace transform, one can obtain

$$P_{up}(t) = 0.01374899899 \, e^{(-1.048649987t)} - 0.01563143352$$
$$e^{(0.9758559581t)} \cos(0.04729411825t) + 0.03410365800$$
$$e^{(-0.08481904821t)} \sin(0.0478410197t) + 0.04841180140$$
$$e^{(-0.08481904821t)} \cos(0.04728410197t) + 0.03504753275$$
$$e^{(0.9758559581t)} \sin(0.04729411825t) + 0.9534706331 \qquad (62)$$

When $p = 0.7$ and $q = 0.3$, i.e. if the possibility of degraded system after failure is replaced by new one is greater and all the other parameters are kept fixed as described above, then availability of the system is given by

$$P_{up}(t) = 0.9607993851 + 0.01596805560 \, e^{(-0.08497152367t)}$$
$$\sin(0.04629652405t) + 0.009540980126 \, e^{(-1.034675510t)}$$
$$+ 0.02545000729 \, e^{(-0.9826907215t)} \sin(0.03441730621t)$$
$$- 0.01200470779 \, e^{(-0.9826907215t)} \cos(0.03441730621t)$$
$$+ 0.04166434260 \, e^{(-0.08497152367t)} \cos(0.04629652405t) \qquad (63)$$

Varying t from 0 to 10 in equations (62) and (63), one can obtain the behavior of availability with respect to time when (i) $p < q$ and (ii) $p > q$. Table 3 and Fig. 4 are obtained corresponding to availability analysis of the system.

Table 3: Time vs. availability

Time	Availability	
	$p < q$	$p > q$
0	1	1
1	0.998934491	0.99893485
2	0.996812166	0.996819273
3	0.994395909	0.99443029
4	0.991936342	0.992031343
5	0.989514737	0.989710054
6	0.987159094	0.987494809
7	0.98488137	0.985394174
8	0.982688563	0.983409935
9	0.980585686	0.981541321
10	0.978576443	0.979786367

8.3 MTTF analysis

The MTTF of the system can be calculated with the help of the formula given below:

$$\text{MTTF} = \lim_{s \to 0} \overline{P}_{up}(s)$$

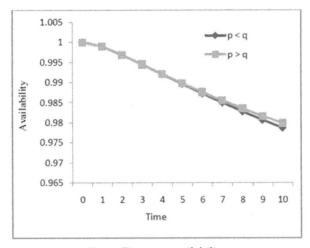

Fig. 4: Time vs. availability

Now setting $u_1 = u_2 = v_1 = v_2 = \Phi = \psi = h = 0$ and varying failure rates we have following two cases:

1. Changing λ_1 from 0.10 to 1.0 and assume $\lambda_2 = 0.2$, $n = 6$, $\theta = 1$, $x = y = z = 1$ in equation (61), one can obtain the variation of MTTF with respect to λ_1.
2. Increasing the value of λ_2 from 0.10 to 1.0 and $\lambda_1 = 0.1$, $n = 6$, $\theta = 1$, $x = y = z = 1$ in equation (61), we get how MTTF changes with respect to λ_2.

Variation of MTTF with respect to λ_1 and λ_2 in the above two cases has been given in Figs 6 and 7, respectively.

Table 4: MTTF with respect to failure rate λ_1

Failure rate (λ_1)	MTTF w. r. t. λ_1
0.1	6.47310206
0.2	3.633481556
0.3	2.528610574
0.4	1.939472963
0.5	1.573112422
0.6	1.323215121
0.7	1.141848864
0.8	1.004216787
0.9	0.896199534
1	0.809165856

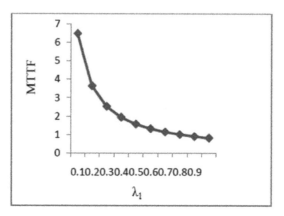

Fig. 5: MTTF with respect to failure rate λ_1

Table 5: MTTF with respect to failure rate λ_2

Failure rate (λ_2)	MTTF w. r. t. λ_2
0.1	7.266963108
0.2	3.878945926
0.3	2.646430242
0.4	2.008433575
0.5	1.618331714
0.6	1.355135601
0.7	1.165578597
0.8	1.02254705
0.9	0.910783607
1	0.821044979

8.4 Cost analysis

If the service facility is always available, then expected profit during the interval $(0, t]$ is given by

$$E_P(t) = K_1 \int_0^t P_{up}(t)dt - K_2 t$$

where K_1 and K_2 are the revenue per unit time and service cost per unit time, respectively. Using the expression of $P_{up}(t)$ given in equation (62), one can obtain

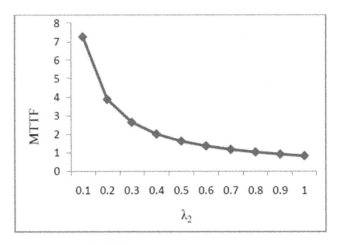

Fig. 6: MTTF with respect to failure rate λ_2

Table 6: Time vs. expected profit when $p < q$

Time	$E_p(t)$		
	$K_2 = 0.1$	$K_2 = 0.2$	$K_2 = 0.3$
0	0	0	0
1	0.899613894	0.799613894	0.699613894
2	1.79753125	1.59753125	1.39753125
3	2.693145225	2.393145225	2.093145225
4	3.586310257	3.186310257	2.786310257
5	4.477031068	3.977031068	3.477031068
6	5.365361869	4.765361869	4.165361869
7	6.251375275	5.551375275	4.851375275
8	7.135152933	6.335152933	5.535152933
9	8.016782391	7.116782391	6.216782391
10	8.89635553	7.89635553	6.89635553

$$
\begin{aligned}
E_p(t) = \ & K_1(-0.01311114210 \, e^{(-1.048649987t)} + 0.01424414403 \, e^{(-0.9758559581t)} \\
& \cos(0.04729411825t) - 0.03660498938 \, e^{(-0.9758559581t)} \\
& \sin(0.04729411825t) - 0.6064442256 \, e^{(-0.08481904821t)} \\
& \cos(0.04728410197t) - 0.06400080539 \, e^{(-0.08481904821t)} \\
& \sin(0.04728410197t) + 0.9534706331t + 0.6053112237) - tK_2 \quad (64)
\end{aligned}
$$

This is the equation of expected profit when the possibility of degraded system after failure is replaced by new one is less, i.e. $p = 0.2$, $q = 0.8$ but if $p = 0.7$, $q = 0.3$, then expected profit is given by

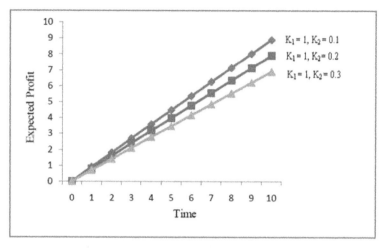

Fig. 7: Time vs. expected profit when $p < q$

$$E_p(t) = K_1(0.9607993851t) - 0.4570444127\ e^{(-0.08497152367t)}$$
$$\cos(0.04629652405t) + 0.06109708076\ e^{(-0.08497152367t)}$$
$$\sin(0.04629652405t) - 0.009221229297\ e^{(-1.034675510t)} +$$
$$0.01129525573\ e^{(-0.9826907215t)} \cos(0.03441730621t)$$
$$- 0.02629388779\ e^{(-0.9826907215t)} \sin(0.03441730621t)$$
$$+ 0.4549703863) - t\,K_2 \tag{65}$$

Keeping $K_1 = 1$ and varying K_2 as 0.1, 0.2, 0.3, 0.4 and 0.5 in equations (64) and (65), one can obtain Table 7 and, correspondingly, Fig. 8.

Table 7: Time vs. expected profit when $p > q$

Time	$E_p(t)$		
	$K_2 = 0.1$	$K_2 = 0.2$	$K_2 = 0.3$
0	0	0	0
1	0.899613958	0.799613958	0.699613958
2	1.797533983	1.597533983	1.397533983
3	2.693166372	2.393166372	2.093166372
4	3.586392946	3.186392946	2.786392946
5	4.477255516	3.977255516	3.477255516
6	5.365848594	4.765848594	4.165848594
7	6.252283403	5.552283403	4.852283403
8	7.136675766	6.336675766	5.536675766
9	8.019141826	7.119141826	6.219141826
10	8.899796302	7.899796302	6.899796302

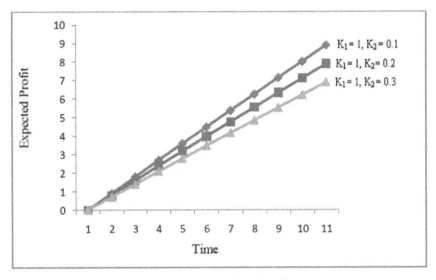

Fig. 8: Time vs. expected profit when $p > q$

8.5 Sensitivity analysis

Putting $u_1 = u_2 = v_1 = v_2 = \Phi = \psi = h = 0$, $\theta = 1$ and $x = y = z = 1$ in equation (58), we have

$$R(s) = \frac{1}{s + \frac{1}{2}n\lambda_1 + \frac{1}{2}n\lambda_2} + \frac{1}{2} \times \frac{n\lambda_1}{\left(s + \frac{1}{2}n\lambda_1 + \frac{1}{2}n\lambda_2\right)\left(s + \left(\frac{1}{2}n - 1\right)\lambda_1 + \frac{1}{2}n\lambda_2\right)}$$

$$+ \frac{1}{2} \times \frac{n\lambda_2}{\left(s + \frac{1}{2}n\lambda_1 + \frac{1}{2}n\lambda_2\right)} \times \frac{1}{\left(s + \left(\frac{1}{2}n - 1\right)\lambda_2 + \frac{1}{2}n\lambda_1\right)}$$

First, we find $\partial R(s)/\partial \lambda_1$ by differentiating equation (66) with respect to λ_1, then we take its inverse Laplace transform. Now, putting $\lambda_2 = 0.02$, $n = 6$ and then varying the time as $t = 0, 1, 2, 3, 4, 5, 6, 7, 8, 9$ and 10 for $\lambda_1 = 0.1, 0.2$ and 0.3, we obtain the sensitivity analysis of system reliability with respect to λ_1, which is demonstrated by Table 8 and Fig. 9.

Similarly, by differentiating the equation (66) with respect to λ_2, we can find $\partial R(s)/\partial \lambda_2$. By finding the inverse Laplace transform of $\partial R(s)/\partial \lambda_2$ and putting $\lambda_1 = 0.01$, $n = 6$ and then varying the time as $t = 0, 1, 2, 3, 4, 5, 6, 7, 8, 9$ and 10 for $\lambda_2 = 0.1, 0.2$ and 0.3, we get the sensitivity analysis of system reliability with respect to λ_2 (depicted by Table 9 and Fig. 10).

Table 8: Sensitivity of system reliability with respect to
different values of λ_1

Time	Value of $\partial R(t) / \partial \lambda_1$		
	$\lambda_1 = 0.1$	$\lambda_1 = 0.2$	$\lambda_1 = 0.3$
0	0	0	0
1	-0.567097524	-0.78056367	-0.873365234
2	-1.650784758	-1.772840844	-1.554036667
3	-2.705572938	-2.273686257	-1.568542397
4	-3.507034824	-2.312912708	-1.261269043
5	-3.999295758	-2.075826935	-0.898551463
6	-4.20718806	-1.723477848	-0.594526965
7	-4.187509211	-1.357573889	-0.374565428
8	-4.003460965	-1.029884746	-0.228033235
9	-3.712469937	-0.759750017	-0.135403407
10	-3.361450623	-0.548595093	-0.078907938

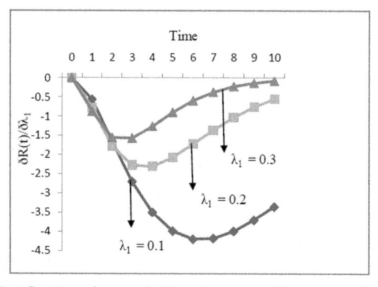

Fig. 9: Sensitivity of system reliability with respect to different values of λ_1

Table 9: Sensitivity of system reliability with respect to different values of λ_2

Time	$\partial R(t)/\partial\lambda_2$		
	$\lambda_2 = 0.1$	$\lambda_2 = 0.2$	$\lambda_2 = 0.3$
0	0	0	0
1	-0.518686956	-0.755677491	-0.863916521
2	-1.56112734	-1.777240115	-1.592383076
3	-2.645526072	-2.359803721	-1.664215676
4	-3.545681688	-2.484767266	-1.385019586
5	-4.180703252	-2.307829327	-1.020776269
6	-4.547372511	-1.982456179	-0.698401783
7	-4.679734156	-1.61525806	-0.45480074
8	-4.625825414	-1.267185902	-0.286071985
9	-4.435001108	-0.966467944	-0.175439638
10	-4.151653873	-0.721316581	-0.105558198

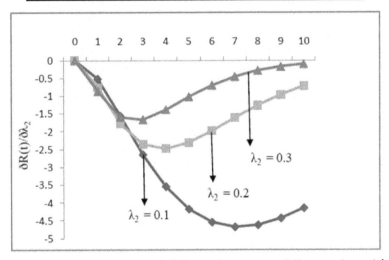

Fig. 10: Sensitivity of system reliability with respect to different values of λ_2

9. Conclusion

For a more concrete study of the system behavior, curves for reliability, availability and expected profit with respect to time, MTTF and sensitivity with respect to λ_1 and λ_2 have been plotted. By critical examination of Fig. 3, we conclude that the reliability of the system decreases as time increases and it attains a value 0.89 at $t = 10$. Figure 4 shows the availability of the system when: (i) possibility of replacement of degraded system after failure by new one is less, i.e. $p < q$ and (ii) $p > q$. It is clear from Fig. 4 that availability of the system decreases in both the cases with the increment in the time. One of the interesting facts obtained is that the availability of the system is more in the case (ii) in comparison to the case (i).

Figures 5 and 6 depict the behavior of MTTF with respect to λ_1 and λ_2 respectively. Observation of these figures reveals that in each case MTTF of the system decreases as the failure rate increases from 0.1 to 1. It varies from 6.47-0.80 and 7.26-0.82 with respect to λ_1 and λ_2 respectively. Also, MTTF of the system is greater with respect to λ_2 than λ_1.

For the cost analysis of the system we keep revenue cost per unit time at 1 and vary service cost from 0.1 to 0.5. The behavior of expected profit can be observed from Figs 7 and 8. The profit goes on decreasing with the increase in service cost. One can draw an important conclusion from Fig. 8. The system becomes more profitable when the chances of degraded system is replaced by new one after failure is greater, i.e. $p > q$. The highest and lowest values of expected profit are obtained to be 8.89635553 and 0.699613894 when $p < q$ and 8.899796302 and 0.699613958 when $p > q$ respectively.

Figures 9 and 10 represent how sensitivity of the system reliability varies with respect to parameters λ_1 and λ_2. It is clear from these figures that sensitivity of system reliability initially decreases and then increases as time passes with respect to λ_1 and λ_2 and increases with the increase of failure rates λ_1 and λ_2 from 0.1 to 0.3. It is interesting to note that system reliability is more sensitive with respect to λ_1. By studying the graphs, we can conclude that the system can be made less sensitive by controlling its failure rates.

Hence, on the basis of the results obtained for availability and expected profit for a particular case, it is concluded that the system can be made profitable if the degraded system is replaced by new one when it fails.

REFERENCES

Chander, S. and M. Singh. 2009. Reliability modeling of 2-out-of-3 redundant system subject to degradation after repair. Journal of Reliability and Statistical Studies 2(2): 91-104.

Chen, Y. and Q. Yang. 2005. Reliability of two-stage weighted-k-out-of-n systems with components in common. IEEE Transactions on Reliability 54(3): 431-440.

Coit, D.W. and J. Liu. 2000. System reliability optimization with k-out-of-n subsystems. International Journal of Reliability, Quality and Safety Engineering 7(2): 129-142.

Dhillon, B.S. 1992. Stochastic modeling of K-out-of-n units family of system. International Journal of Systems Sciences 28(8): 1277–1287.

Kumar, D. and S.B. Singh. 2016. Stochastic analysis of complex repairable system with deliberate failure emphasizing reboot delay. Communications in Statistics – Simulation and Computation 45(2): 583–602.

Kumar, J., M.S. Kadyan and S.C. Malik. 2010. Cost-benefit analysis of a two-unit parallel system subject to degradation after repair. Applied Mathematical Sciences 4(56): 2749–2758.

Malik, S.C., M.S. Kadyan and J. Kumar. 2010. Stochastic analysis of a two-unit parallel system subject to degradation and inspection for feasibility of repair. Journal of Mathematics and System Sciences 6(1): 5–13.

Malik, S.C., P. Chand and J. Singh. 2008. Stochastic analysis of an operating system with two types of inspection subject to degradation. Journal of Applied Probability and Statistics (USA) 3(2): 227–241.

Manglik, M. and M. Ram. 2015. Behavioural analysis of a hydroelectric production power plant under reworking scheme. International Journal of Production Research 53(2): 648–664.

Mokaddis, G.S., S.W. Labib and A.M. Ahmed. 1997. Analysis of a two-unit warm standby system subject to degradation. Microelectronics and Reliability 37(4): 641–647.

Murari, K. and V. Goyal. 1984. Comparison of two-unit cold standby reliability models with three types of repair facilities. Microelectronics and Reliability 24(1): 35–49.

Nailwal, B. and S.B. Singh. 2011. Performance evaluation and reliability analysis of a complex system with three possibilities in repair with the application of copula. International Journal of Reliability and Application 12(1): 15–39.

Nailwal, B. and S.B. Singh. 2012a. Reliability measures and sensitivity analysis of a complex matrix system including power failure. International Journal of Engineering 25(1): 115–130.

Nailwal, B. and S.B. Singh. 2012b. Reliability and sensitivity analysis of an operating system with inspection in different weather conditions. International Journal of Reliability, Quality and Safety Engineering 19(2). DOI: 10.1142/S021853931250009X

Nailwal, B. and S.B. Singh. 2017. Reliability analysis of two dissimilar-cold standby redundant systems subject to inspection with preventive maintenance using copula. IGI Global 201–221. DOI: 10.4018/978-1-5225-1639-2.ch010

Nelson, R.B. 2006. An Introduction to Copulas (2nd ed.). New York: Springer.

Promila, S.C. Malik and M.S. Kadyan. 2010. Analysis of an operating system with inspection subject to degradation and different weather conditions. International Journal of Engineering Science and Technology 2(11): 6779–6787.

Ram, M. and A. Kumar. 2014a. Performance of a structure consisting a 2-out-of-3:F substructure under human failure. Arab. J. Sci. Eng. 39: 8383–8394

Ram, M. and A. Kumar. 2014b. Performability analysis of a system under 1-out-of-2:G scheme with perfect reworking. J. Braz. Soc. Mech. Sci. Eng. 37(3): 1029–1038.

Ram, M., S.B. Singh and V.V. Singh. 2013. Stochastic analysis of a standby system with waiting repair strategy. IEEE Transactions on Systems, Man, and Cybernetics: Systems 43(3): 698–707.

Sen, P.K. 2003. Copulas: concepts and novel applications. International Journal of Statistics LXI (3): 323–353.

Singh, S.K. 1989. Profit evaluation of a two-unit cold standby system with random appearance and disappearance of the service facility. Microelectronics and Reliability 29: 705–709.

Srinivasan, S.K. and M.N. Gopalan. 1973. Probabilistic analysis of a two-unit system with warm standby and single repair facility. Operations Research 21: 740–754.

Wu, J.-S. and R.-J. Chen. 1994. An algorithm for computing the reliability of weighted-k-out-of-n systems. IEEE Transactions on Reliability 43(2): 327–328.

Yam, C.M. Richard, Zuo J. Ming, Y.L. Zhang and L. Yuan. 2003. A method for evaluation of reliability indices for repairable circular consecutive-k-out-of-n:F systems. Reliability Engineering and System Safety 79: 1–9.

Signature of Linear Consecutive *k*-out-of-*n* System

Akshay Kumar[1*] and Mangey Ram[2]

[1] Department of Applied Sciences, Tula's Institute, The Engineering and Management College, Dehradun, Uttarakhand, India

[2] Department of Mathematics, Computer Science & Engineering, Graphic Era (Deemed to be University), Dehradun, Uttarakhand, India

1. Introduction

Reliability theory is a key tool for engineering systems and their application in various fields. In previous years, many researchers computed reliability of different systems, including basic systems such as series and parallel; k-out-of-n system is the special case of series parallel system. If n components are working, it is called a series system, and if one component is working, it is called a parallel system. Chiang and Niu (1981) considered a consecutive k-out-of-n system and computed the exact reliability of the system with upper and lower form and found that if k consecutive components fail, the system also fails. Hwang (1982) studied the consecutive k-out-of-n:F system; when at least k consecutive components have failed then the system has also failed. A solution to the proposed system was defined in the case of reliability analysis. Fu (1985) computed the system reliability for large consecutive k-out-of-n:F with the help of an equal number of failure components. Zhang (1988) studied the different cases of linear consecutive k-out-of-n:G systems and discussed the reliability function and mean time to failure. Kossow and Preuss (1989) proposed a consecutive k-out-of-n:F system with n components in sequence and evaluated the reliability in linear and circular cases.

*Corresponding author: akshaykr1001@gmail.com

Nomenclature

R/H = Reliability/Structure function of the system

$E(T)$ = expected lifetime of the system

$E(X)$ = expected value of X

   C_j = Minimal signature of the system

   b_j = Signature of the system

   B_a = Tail signature of the system

Rushdi (1993) discussed the characteristics of k-out-of-n system with the help of generating function and based on inclusion-exclusion technique. Preuss and Boehme (1994) assessed the properties of a linear and circular consecutive k-out-of-n system and described its series and parallel systems when components are 1 and k. Eryilmaz (2009) discussed the properties of a consecutive k-out-of-n system and evaluated the mean time to failure, failure rate and failure mean residual lifetime of the proposed system. Ram and Singh (2010) discussed the reliability, availability, mean time to failure and cost analysis of the proposed complex system with the help of copula technique. Salehi et al. (2011) proposed a consecutive k-out-of-n system with both linear and circular forms and discussed the characterised properties of residual lifetime components when at least k component are working. Song et al. (2012) analysed the reliability and maintenance of k-out-of-n redundancy system from supplementary techniques and optimized the reliability and cost.

Further, Ram (2013) studied a survey on system reliability approaches. In that study he surveyed the different methods and techniques for computing reliability in various engineering fields, such as Bayes approach, coherent systems approaches, copula approach, coverage factor approach, designed experiments, distributed system, genetic algorithm, network system approach, etc. Salehi (2016) discussed the reliability and residual lifetime of the proposed consecutive k-out-of-n system. A linear and circular consecutive k-out-of-n system works if k consecutive components are working. Kumar et al. (2016) evaluated the interval-valued reliability of 2-out-of-4 system using universal generating function when the system works if at least 2 components out of 4 work.

Further, signature plays a key role in the field of system reliability theory and also in analysing the mean time to failure, sensitivity and cost. Eryilmaz (2010) computed the signature and expected value of a consecutive k-out-of-n complex system using distribution of ordered statistics while components are in a working state. Eryilmaz et al. (2011) studied the signature of an m-consecutive k-out-of-n:F system having exchangeable elements. They analysed the signature reliability, expected lifetime and cost analysis using ordered statistics and IFR prevention property. Triantafyllou and Koutras (2011) provided the signature-based computation of a 2-within-consecutive k-out-of-n:F system and evaluated the signature reliability and expected lifetime with the help of ordered statistics and stochastic comparison. Eryilmaz (2012a) proposed m-consecutive-k-out-of-n:F system and evaluated the signature reliability analysis and discussed its applications in various fields, such as telecommunication systems, heating systems and oil pipeline systems, etc. Da and Hu (2013) discussed the properties of a coherent system with signature and computed the signature of 3-state system on bivariate system having independent modules. Da et al. (2014) discussed the characteristics of k-out-of-n system having n modules and demonstrated the signature reliability with minimal signature. Kumar and Singh (2017a, 2017b, 2018) analysed the signature reliability, expected cost rate, sensitivity and expected lifetime of a sliding window system, linear multi-state sliding window system and complex system by using the universal generating function (UGF) method and various algorithms.

In the above discussion, many researchers computed the reliability of the consecutive and linear consecutive k-out-of-n system, as well as many more systems in previous years. In this study, the signature and its parameters, such as tail signature, expected lifetime and expected cost, have been evaluated with the help of reliability functions.

2. Model description

The linear consecutive k-out-of-n system is a type of k-out-of-n system. The system consists of n components arranged in linear configuration and it works if at least k consecutive components are working, as shown in Fig. 1.

$$x = \begin{cases} 1, \text{ if } i^{\text{th}} \text{ element is working} \\ 0, \text{ if } i^{\text{th}} \text{ elements is faild} \end{cases}.$$

If all components of the system have equal reliability, i.e. $R_1 = R_2 = R_3 = \ldots R_n = R$, ,then the reliability of the system can be computed as $R(n; k) = R(n-1; k) + ur^k (1 - R(n-k-1; k))$ (Zhang 1998).

Now, reliability function R of the proposed system is obtained, having i.i.d. components (Boland, 2001) as

$$R = \frac{1}{\binom{m}{m-n+1}} \sum_{a \subseteq |m|} \phi(R(n;k)) - \frac{1}{\binom{m}{m-a}} \sum \phi(R(n;k))$$

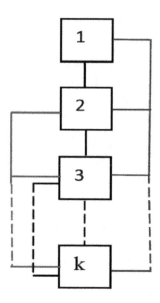

Fig. 1: *k*-out-of-*n* linear consecutive system

3. Algorithm to compute the signature of a linear consecutive *k*-out-of-*n* system

Step 1: Find the signature of the structure function (Boland, 2001).

$$B_a = \frac{1}{\binom{m}{m-n+1}} \sum_{\substack{H \subseteq |m| \\ |H| = m-n+1}} \varphi(H) - \frac{1}{\binom{m}{m-n}} \sum_{\substack{H \subseteq |m| \\ |H| = m-n}} \varphi(H) \tag{1}$$

Compute polynomial function $H(P)$ of linear consecutive *k*-out-of-*n* system by

$$H(P) = \sum_{e=1}^{m} C_e \binom{m}{e} P^e q^{n-e}, \text{ where, } C_j = \sum_{j=m-e+1}^{m} B_j, e = 1, 2, .., m.$$

Step 2: Calculate the tail signature of system, i.e. $(m + 1)$ – tuple $B_0, ..., B_m$, by using

$$B_a = \sum_{j=a+1}^{m} b_j = \frac{1}{\binom{m}{m-a}} \sum_{|H| = m-a} \varphi(H) \tag{2}$$

Step 3: Find the structure function in the form of a polynomial with the help of Taylor expansion as follows:

$$P(x) = y^m h\left(\frac{1}{y}\right) \tag{3}$$

Step 4: Determine the tail signature of the system using equation (1) (Marichal and Mathonet, 2012).

$$B_a = \frac{(m-a)!}{m!} D^a P(1), \, a = 0, ..., m \tag{4}$$

Step 5: Determine the signature of the system using step 3 as follows:

$$b = B_{a-1} - B_a, \, a = 1, ..., m. \tag{5}$$

3.1 Algorithm to obtain the expected lifetime of a system having minimum signature

Step 1: Obtain $E(T)$ of system of i.i.d. components by (see Navarro and Rubio, 2009).

$$E(T) = \mu \sum_{j=1}^{n} \frac{C_j}{j} \tag{6}$$

where, $C_j (j = 1, ..., n)$ is a vector coefficient of minimal signature.

3.2 Algorithm for evaluating the expected value of the component X and expected cost rate of a linear consecutive k-out-of-n system when working elements have failed

Step 1: Calculate the number of failed elements at the time of system failure with signature (Eryilmaz, 2012b).

$$E(X) = \sum_{j=1}^{n} j.b_j, \, j = 1, 2, ..., n \tag{7}$$

Step 2: Compute the $E(X)$ and $E(X)/E(T)$ of the system with minimum signature.

4. Example

Consider a 2-out-of-5 linear consecutive system with components having uniform reliability, as shown in Fig. 2.

Signature of the proposed system is defined as:

$$R(n; k) = R(n-1; k) + ur^k (1 - R(n-k-1; k)) \text{ (Zhang, 1998)} \tag{8}$$

For $n = 0, 1, ..., 5$ putting in equation (1), we have

$$R(0, 2) = R(-1, 2) + ur^2 (1 - R(-3, 2)) = 0.$$

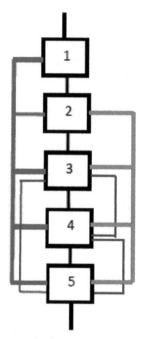

Fig. 2: 2-out-of-5 linear consecutive system

$$R(1, 2) = R(0, 2) + ur^2(1 - R(-2, 2) = 0 \cdot$$

$$R(2, 2) = r^2 \cdot$$

$$R(3, 2) = R(2, 2) + ur^2(1 - R(0, 2) = r^2 + (1 - r)r^2 \cdot$$

$$R(4, 2) = R(3, 2) + ur^2(1 - R(1, 2) = r^2 + (1 - r)r^2 + (1 - r)r^2 \cdot$$

$$R(5, 2) = R(4, 2) + ur^2(1 - R(2, 2) = 3r^2 - 2r^3 + (1 - r)r^2(1 - r^2) \cdot$$

Hence, the reliability function is

$$R = 4r^2 - 3r^3 - r^4 + r^5 \tag{9}$$

5. Signature of linear consecutive 2-out-of-5 system

Using equation (3), we get the polynomial function of the system

$$R = 4r^2 - 3r^3 - r^4 + r^5 \tag{10}$$

From using equation (10), we get the tail signature using algorithm 4.1 and step 4

$$B = \left(1, 1, \frac{9}{10}, \frac{2}{5}, 0, 0\right).$$

Using step 3, compute the signature of the system as follows:

$$b = \left(0, \frac{1}{10}, \frac{5}{10}, \frac{2}{5}, 0\right).$$

5.1 Expected lifetime of linear consecutive 2-out-of-5 system

Using equation (9), we get minimal signature

$$R = 4r^2 - 3r^3 - r^4 + r^5.$$

Minimal signature = (0, 4, –3, –1, 1).

Hence, expected lifetime is

$$E(T) = 0.95.$$

5.2 Expected cost rate of a linear consecutive 2-out-of-5 system

By using algorithm 4.3 we get the expected value of X, denoted as $E(X)$

$$E(X) = 3.3$$

Cost rate = (0.347368).

6. Conclusion

In this study we have discussed the signature of a linear consecutive 2-out-of-5 system and evaluated some parameters, such as tail signature, signature, expected lifetime and expected cost using reliability functions. The signature of the system $\left(0, \frac{1}{10}, \frac{5}{10}, \frac{2}{5}, 0\right)$ increases, expected lifetime is (0.95) and expected cost is (0.347368).

REFERENCES

Boland, P.J. 2001. Signatures of indirect majority systems. Journal of Applied Probability 38(2): 597–603.

Chiang, D.T. and S.C. Niu. 1981. Reliability of consecutive-*k*-out-of-*n*:F system. IEEE Transactions on Reliability 30(1): 87–89.

Da, G. and T. Hu. 2013. On bivariate signatures for systems with independent modules. pp. 143–166. *In*: Stochastic Orders in Reliability and Risk. Springer, New York, NY.

Da, G., L. Xia and T. Hu. 2014. On computing signatures of *k*-out-of-*n* systems consisting of modules. Methodology and Computing in Applied Probability 16(1): 223–233.

Eryılmaz, S. 2009. Reliability properties of consecutive *k*-out-of-*n* systems of arbitrarily dependent components. Reliability Engineering & System Safety 94(2): 350–356.

Eryilmaz, S. 2010. Number of working components in consecutive *k*-out-of-*n* system while it is working. Communications in Statistics Simulation and Computation 39(4): 683–692.

Eryilmaz, S. 2012a. *m*-consecutive-*k*-out-of-*n*:F system with overlapping runs: signature-based reliability analysis. International Journal of Operational Research 15(1): 64–73.

Eryilmaz, S. 2012b. The number of failed components in a coherent system with exchangeable components. IEEE Transactions on Reliability 61(1): 203–207.

Eryilmaz, S., M.V. Koutras and I.S. Triantafyllou. 2011. Signature-based analysis of *m*-Consecutive-*k*-out-of-*n*:F systems with exchangeable components. Naval Research Logistics (NRL) 58(4): 344–354.

Fu, J.C. 1985. Reliability of a large consecutive-*k*-out-of-*n*:F system. IEEE Transactions on Reliability 34(2): 127–130.

Hwang, F.K. 1982. Fast solutions for consecutive-*k*-out-of-*n*:F system. IEEE Transactions on Reliability 31(5): 447–448.

Kossow, A. and W. Preuss. 1989. Reliability of consecutive-*k*-out-of-*n*:F systems with nonidentical component reliabilities. IEEE Transactions on Reliability 38(2): 229–233.

Kumar, A. and S.B. Singh. 2017a. Computations of signature reliability of coherent system. International Journal of Quality & Reliability Management 34(6): 785–797.

Kumar, A. and S.B. Singh. 2017b. Signature reliability of sliding window coherent system. pp. 83–95. *In:* M. Ram and J.P. Davim (Eds). Mathematics Applied to Engineering. Elsevier Publisher.

Kumar, A. and S.B. Singh. 2018. Signature reliability of linear multi-state sliding window system. International Journal of Quality & Reliability Management 35(10): 2403–2413.

Kumar, A., S.B. Singh and M. Ram. November 2016. Interval-valued reliability assessment of 2-out-of-4 system. pp. 1–4. *In:* Emerging Trends in Communication Technologies (ETCT), International Conference on IEEE.

Marichal, J.L. and P. Mathonet. 2013. Computing system signatures through reliability functions. Statistics & Probability Letters 83(3): 710–717.

Navarro, J. and R. Rubio. 2009. Computations of signatures of coherent systems with five components. Communications in Statistics-Simulation and Computation 39(1): 68–84.

Preuss, W.W. and T.K. Boehme. 1994. On reliability analysis of consecutive-*k* out-of-*n*:F systems and their generalizations: a survey. pp. 401–411. *In:* Approximation, Probability, and Related Fields. Springer, Boston, MA.

Ram, M. 2013. On system reliability approaches: Aa brief survey. International Journal of System Assurance Engineering and Management 4(2): 101–117.

Ram, M. and S.B. Singh. 2010. Availability, MTTF and cost analysis of complex system under preemptive-repeat repair discipline using Gumbel-Hougaard family copula. International Journal of Quality & Reliability Management 27(5): 576–595.

Rushdi, A.M. 1993. Reliability of *k*-out-of-*n* systems. pp. 185–227. *In*: Fundamental Studies in Engineering (Vol. 16). Elsevier.

Salehi, E. 2016. On reliability analysis of consecutive *k*-out-of-*n* systems with arbitrarily dependent components. Applications of Mathematics 61(5): 565–584.

Salehi, E.T., M. Asadi and S. Eryılmaz. 2011. Reliability analysis of consecutive k-out-of-n systems with non-identical components lifetimes. Journal of Statistical Planning and Inference 141(8): 2920–2932.

Song, S., N. Chatwattanasiri, D.W. Coit, Q. Feng and N. Wattanapongsakorn. 2012. Reliability analysis for *k*-out-of-*n* systems subject to multiple dependent competing failure processes. 2012 International Conference on Quality, Reliability, Risk, Maintenance, and Safety Engineering. pp. 36–42. Chengdu.

Triantafyllou, I.S. and M.V. Koutras. 2011. Signature and IFR preservation of 2-within-consecutive *k*-out-of-*n*:F systems. IEEE Transactions on Reliability 60(1): 315–322.

Zhang, W. 1988. Theory and analysis of consecutive-*k*-out-of-*n*:G systems reliability. Retrospective Theses and Dissertations. 9749. https://lib.dr.iastate.edu/rtd/9749.

Index

Milton Keynes UK
Ingram Content Group UK Ltd.
UKHW040101071024
449327UK00019B/707